Craniofacial Identification in Forensic Science

Craniofacial Identification in Forensic Science

Er. Devesh Kumar Dixit

RANDOM

RANDOM PUBLICATIONS

NEW DELHI (INDIA)

Craniofacial Identification in Forensic Science

ISBN 978-93-5111-252-5

Published in 2024 in India by

RANDOM PUBLICATIONS

4376-A/4B, Gali Murari Lal, Ansari Road
New Delhi-110 002
Phone: +9111-43580356, 23289044
e-mail: sales@randompublications.com,
info@randompublications.com, randomexports@gmail.com

Type Setting by : Keystoneprintads, Delhi-110051
Digitally Printed at : Replika Press Pvt. Ltd.

Preface

Forensic science is the scientific method of gathering and examining evidence. This is especially important in law enforcement where forensics is done in relation to criminal or civil law, but forensics are also carried out in other fields, such as astronomy, archaeology, biology and geology to investigate ancient times.

The face provides most people with information about a person's identity. Similarly, in forensic science it has long been recognized that the most useful methods of obtaining the identity of an individual, in the absence of fingerprints, rely on the characteristics of the human face, head and teeth. Post-mortem identification of skeletal remains depends heavily on dentition and the morphology of the skull.

Forensic facial reconstruction (or forensic facial approximation) is the process of recreating the face of an individual (whose identity is often not known) from their skeletal remains through an amalgamation of artistry, forensic science, anthropology, osteology, and anatomy. It is easily the most subjective—as well as one of the most controversial—techniques in the field of forensic anthropology. Despite this controversy, facial reconstruction has proved successful frequently enough that research and methodological developments continue to be advanced.

Craniofacial superimposition is a forensic process in which a photograph of a missing person is compared with a skull found to determine its identity. After one century of development, craniofacial superimposition has become an interdisciplinary research field where computer sciences have acquired a key role as a complement of forensic sciences. Moreover, the availability of new digital equipment (such as computers and 3D scanners) has resulted in a significant advance in the applicability of this forensic identification technique. The purpose of this contribution is twofold. On the one hand, we aim to clearly define the different stages involved in the computer-aided craniofacial superimposition process.

Some forensic scientists confuse their role with that of an attorney. The attorney's role is to present the best one-sided case for his/her client, the only exception being not to argue for anything the attorney knows to be untrue.

Unlike an expert witness, attorneys take no oath in court. In contrast, forensic scientists take an oath 'to tell the truth, the whole truth, and nothing but the truth'. If forensic scientists assume the attorney's total advocacy role and present the best one-sided case for the retaining attorney regardless of the 'truth' and the oath taken, the forensic experts may be perceived as 'hired guns'. They thereby vilify not only their own reputation, but also taint that of the entire field. Many forensic scientists consider the 'hired gun' problem the most serious ethical problem in forensic science.

This compendium is aimed to help the criminal justice system in the dissemination of justice through the aids of forensic science and the students and teachers in the universities and medical colleges in their studies and research in the field of forensic science.

I thank all members of my team who have helped in the preparation of the book. My special thanks go to "Random Publications" who have published the book.

– Er. Devesh Kumar Dixit

Contents

1

Forensic Science

INTRODUCTION

To put forensic science in simply layman's terms: it is the application of scientific procedures in conjunction with the law. Forensic science is the umbrella term for a collection of scientific procedures that can be employed when it comes to providing evidence that a crime has been committed and also provides ways and means of proving how a crime has been committed. In its simplest form it can be a means by which fingerprints, fibres, DNA and bodily fluids can be collected and stored - without contamination - so that they can be later used as evidence in a police investigation into a specific crime - or indeed series of crimes.

Forensic science (often known as forensics) is the scientific method of gathering and examining evidence. This is especially important in law enforcement where forensics is done in relation to criminal or civil law, but forensics are also carried out in other fields, such as astronomy, archaeology, biology and geology to investigate ancient times. n the United States there are over 12,000 Forensic Science technicians, as of 2010.

The word forensic comes from the Latin forensis, meaning "of or before the forum." In Roman times, a criminal charge meant presenting the case before a group of public individuals in the forum.

Both the person accused of the crime and the accuser would give speeches based on their sides of the story. The individual with the best argument and delivery would determine the outcome of the case. This origin is the source of the two modern usages of the word forensic – as a form of legal evidence and as a category of public presentation. Seizure In modern use, the term "forensics" in the place of "forensic science" can be considered correct as the term "forensic" is effectively asynonym for "legal" or "related to courts".

However, the term is now so closely associated with the scientific field that many dictionaries include the meaning that equates the word "forensics" with "forensic science".

HISTORY

Antiquity and the Middle Age

The ancient world lacked standardized forensic practices, which aided criminals in escaping punishment. Criminal investigations and trials relied on forced confessions and witness testimony. However ancient sources contain several accounts of techniques that foreshadow the concepts of forensic science that is developed centuries later, such as the "Eureka" legend told of Archimedes (287–212 BC). The account about Archimedes tells of how he invented a method for determining the volume of an object with an irregular shape. According toVitruvius, a votive crown for a temple had been made for King Hiero II, who had supplied the pure gold to be used, and Archimedes was asked to determine whether some silver had been substituted by the dishonest goldsmith. Archimedes had to solve the problem without damaging the crown, so he could not melt it down into a regularly shaped body in order to calculate its density.

The first written account of using medicine and entomology to solve (separate) criminal cases is attributed to the book of Xi Yuan Lu(translated as Washing Away of Wrongs), written in Song Dynasty China by Song Ci (??, 1186–1249) in 1248. In one of the accounts, the case of a person murdered with a sickle was solved by a death investigator who instructed everyone to bring his sickle to one location. (He realized it was a sickle by testing various blades on an animal carcass and comparing the wound.) Flies, attracted by the smell of blood, eventually gathered on a single sickle. In light of this, the murderer confessed. The book also offered advice on how to distinguish between a drowning (water in the lungs) and strangulation (broken neck cartilage), along with other evidence from examining corpses on determining if a death was caused by murder, suicide or an accident.

Methods from around the world involved saliva and examination of the mouth and tongue to determine innocence or guilt. In ancientChinese cultures, sometimes suspects were made to fill their mouths with dried rice and spit it back out. In ancient middle-eastern cultures the accused were made to lick hot metal rods briefly. Both of these tests had some validity since a guilty person would produce less saliva and thus have a drier mouth. The accused were considered guilty if rice was sticking to their mouth in abundance or if their tongues were severely burned due to lack of shielding from saliva.

Modern history

In the 16th-century Europe medical practitioners in army and university settings began to gather information on cause and manner of death. Ambroise Paré, a French army surgeon, systematically studied the effects of violent death on internal organs. Two Italian surgeons,Fortunato Fidelis and Paolo Zacchia, laid the foundation of modern pathology by studying changes that occurred in the structure of the body as the result of disease. In the late 18th century, writings on these topics began to appear. These included A Treatise on Forensic Medicine and Public Health by the French physician Fodéré and The Complete System of Police Medicine by the German medical expertJohann Peter Franck.

In 1773 a Swedish chemist Carl Wilhelm Scheele devised a way of detecting arsenous oxide, simple arsenic, in corpses, although only in large quantities. This investigation was expanded, in 1806, by German chemist Valentin Ross, who learned to detect the poison in the walls of a victim's stomach, and by English chemist James Marsh, who used chemical processes to confirm arsenic as the cause of death in an 1836 murder trial.

Two early examples of English forensic science in individual legal proceedings demonstrate the increasing use of logic and procedure in criminal investigations. In 1784, in Lancaster, John Toms was tried and convicted for murdering Edward Culshaw with a pistol. When the dead body of Culshaw was examined, a pistol wad (crushed paper used to secure powder and balls in the muzzle) found in his head wound matched perfectly with a torn newspaper found in Toms' pocket. In Warwick in 1816, a farm labourer was tried and convicted of the murder of a young maidservant. She had been drowned in a shallow pool and bore the marks of violent assault. The police found footprints and an impression from corduroy cloth with a sewn patch in the damp earth near the pool.

There were also scattered grains ofwheat and chaff. The breeches of a farm labourer who had been threshing wheat nearby were examined and corresponded exactly to the impression in the earth near the pool. Police started using fingerprints for evidence when Juan Vucetich solved a murder case in Argentina by cutting off a piece of door with a bloody fingerprint on it. Later in the 20th century several British pathologists, Bernard Spilsbury, Francis Camps, Sydney Smith and Keith Simpson pioneered new forensic science methods in Britain. In 1909, Archibald Reissfounded the Institut de police scientifique of the University of Lausanne (UNIL), the first school of forensic science in the world.

Forensic science has been fostered by a number of national forensic science learned bodies including the American Academy of Forensic Sciences (founded 1948), publishers of the Journal of Forensic Sciences; the Canadian Society of Forensic Science (founded 1953), publishers of the Journal of the Canadian Society of Forensic Science; the British Academy of Forensic Sciences (founded 1960), publishers of Medicine, science and the law, and the Australian Academy of Forensic Sciences (founded 1967), publishers of the Australian Journal of Forensic Sciences.

SUBDIVISIONS

- Computational forensics concerns the development of algorithms and software to assist forensic examination.

- Criminalistics is the application of various sciences to answer questions relating to examination and comparison of biological evidence, trace evidence, impression evidence (such as fingerprints, footwear impressions, and tire tracks), controlled substances, ballistics, firearm and toolmark examination, and other evidence in criminal investigations. In typical circumstances evidence is processed in a Crime lab.
- Digital forensics is the application of proven scientific methods and techniques in order to recover data from electronic / digital media. Digital Forensic specialists work in the field as well as in the lab.
- Forensic accounting is the study and interpretation of accounting evidence
- Forensic aerial photography is the study and interpretation of aerial photographic evidence
- Forensic anthropology is the application of physical anthropology in a legal setting, usually for the recovery and identification ofskeletonized human remains.
- Forensic archaeology is the application of a combination of archaeological techniques and forensic science, typically in law enforcement.
- Forensic astronomy uses methods from astronomy to determine past celestial constellations for forensic purposes.
- Forensic botany is the study of plant life in order to gain information regarding possible crimes.
- Forensic chemistry is the study of detection and identification of illicit drugs, accelerants used in arson cases, explosive andgunshot residue.
- Forensic dactyloscopy is the study of fingerprints.
- Forensic document examination or questioned document examination answers questions about a disputed document using a variety of scientific processes and methods. Many examinations involve a comparison of the questioned document, or components of the document, with a set of known standards. The most common type of examination involves handwriting, whereby the examiner tries to address concerns about potential authorship.
- Forensic DNA analysis takes advantage of the uniqueness of an individual's DNA to answer forensic questions such aspaternity/ maternity testing and placing a suspect at a crime scene, e.g. in a rape investigation.
- Forensic engineering is the scientific examination and analysis of structures and products relating to their failure or cause of damage.
- Forensic entomology deals with the examination of insects in, on and around human remains to assist in determination of time or

location of death. It is also possible to determine if the body was moved after death using entomology.

- Forensic geology deals with trace evidence in the form of soils, minerals and petroleum.
- Forensic geophysics is the application of geophysical techniques such as radar for detecting objects hidden underground or underwater.
- Forensic intelligence process starts with the collection of data and ends with the integration of results within into the analysis of crimes under investigation
- Forensic Interviews are conducted using the science of professionally using expertise to conduct a variety of investigative interviews with victims, witnesses, suspects or other sources to determine the facts regarding suspicions, allegations or specific incidents in either public or private sector settings.
- Forensic limnology is the analysis of evidence collected from crime scenes in or around fresh-water sources. Examination of biological organisms, in particular diatoms, can be useful in connecting suspects with victims.
- Forensic linguistics deals with issues in the legal system that requires linguistic expertise.
- Forensic meteorology is a site-specific analysis of past weather conditions for a point of loss.
- Forensic odontology is the study of the uniqueness of dentition, better known as the study of teeth.
- Forensic optometry is the study of glasses and other eye wear relating to crime scenes and criminal investigations
- Forensic pathology is a field in which the principles of medicine and pathology are applied to determine a cause of death or injury in the context of a legal inquiry.
- Forensic podiatry is an application of the study of feet footprint or footwear and their traces to analyze scene of crime and to establish personal identity in forensic examinations.
- Forensic psychiatry is a specialized branch of psychiatry as applied to and based on scientific criminology.
- Forensic psychology is the study of the mind of an individual, using forensic methods. Usually it determines the circumstances behind a criminal's behavior.
- Forensic seismology is the study of techniques to distinguish the seismic signals generated by underground nuclear explosions from those generated by earthquakes.
- Forensic serology is the study of the body fluids.
- Forensic toxicology is the study of the effect of drugs and poisons on/in the human body.

- Forensic video analysis is the scientific examination, comparison and evaluation of video in legal matters.
- Mobile device forensics is the scientific examination and evaluation of evidence found in mobile phones, e.g. Call History and Deleted SMS, and includes SIM Card Forensics
- Trace evidence analysis is the analysis and comparison of trace evidence including glass, paint, fibres and hair.
- Wildlife Forensic Science applies a range of scientific disciplines to legal cases involving non-human biological evidence, to solve crimes such as poaching, animal abuse, and trade in endangered species.

Blood Spatter Analysis is the scientific examination of blood spatter patterns found at a crime scene to reconstruct the events of the crime.

- Forensic Investigation also known as forensic audit is the examination of documents and the interviewing of people to extract evidence.

LITIGATION SCIENCE

Litigation science describes analysis or data developed or produced expressly for use in a trial versus those produced in the course of independent research. This distinction was made by the US 9th Circuit Court of Appeals when evaluating the admissibility of experts. This uses demonstrative evidence, which is evidence created in preparation of trial by attorneys or paralegals.

Examples in Popular Culture

The Argentinean writer Jorge Luis Borges claims that the police novel genre is inaugurated with Edgar Allan Poe's short story, "The Murders in the Rue Morgue". But it is first Sherlock Holmes, the fictional character created by Sir Arthur Conan Doyle in works produced from 1887 to 1915, who used forensic science as one of his investigating methods. Conan Doyle credited the inspiration for Holmes on his teacher at the medical school of the University of Edinburgh, the gifted surgeon and forensic detective Joseph Bell. Agatha Christie's Hercule Poirot and Miss Marple books and television series glorify too a similar prototype.

Decades later the comic strip Dick Tracy also featured a detective using a considerable number of forensic methods, although sometimes the methods were more fanciful than actually possible. Barry Allen (alter ego of The Flash) is a forensic scientist for the Central City police department. Defence attorney Perry Mason occasionally used forensic techniques, both in the novels and television series.

One of the earliest television series to focus on the scientific analysis of evidence was Quincy, M.E. (1976–83, and based loosely on an even earlier Canadian series titled Wojeck), with the title character, a medical examiner working in Los Angeles solving crimes through careful study. The opening theme of each episode featured a clip of the title character, played by Jack Klugman, beginning a lecture to a group of police officers with "Gentlemen, you are about to enter the most fascinating sphere of police work, the world of forensic medicine." Later series with similar premises include Dexter, The Mentalist, CSI, Hawaii Five-0, Cold Case, Bones, Law & Order, Body of Proof, NCIS, Criminal Minds, Silent Witness, Case Closed, Midsomer Murders and Waking the Dead, depict glamorized versions of the activities of 21st-century forensic scientists. Some claim these TV shows have changed individuals' expectations of forensic science, an influence termed the "CSI effect".

Non-fiction TV shows such as Forensic Files, The New Detectives, American Justice, and Dayle Hinman's Body of Evidence have also popularized forensic science. The Ace Attorney series features forensic science, mainly in Apollo Justice: Ace Attorney and the DS-only case in Phoenix Wright: Ace Attorney.

Controversies

Questions about forensic science, fingerprint evidence and the assumption behind these disciplines have been brought to light in some publications, including in theNew York Post. The article stated that "No one has proved even the basic assumption: That everyone's fingerprint is unique." The article also stated that "Now such assumptions are being questioned - and with it may come a radical change in how forensic science is used by police departments and prosecutors." Law professor Jessica Gabel said on NOVA that forensic science, "lacks the rigors, the standards, the quality controls and procedures that we find, usually, in science."

On 25 June 2009 the Supreme Court issued a 5-to-4 decision in Melendez-Diaz v. Massachusetts stating that crime laboratory reports may not be used against criminal defendants at trial unless the analysts responsible for creating them give testimony and subject themselves to cross-examination. The Supreme Court cited the National Academies report Strengthening Forensic Science in the United States in their decision. Writing for the majority, Justice Antonin Scalia referred to the National Research Council report in his assertion that "Forensic evidence is not uniquely immune from the risk of manipulation."

In 2009, scientists indicated that it is possible to fabricate DNA evidence therefore suggesting it is possible to falsely accuse or acquit a person or persons using forged evidence. Although forensic science has greatly enhanced investigators ability to solve crimes, they have limitations and must be scrutinized in and out of the courtroom to avoid wrongful convictions, which have happened.

TYPES OF FORENSIC SCIENCE

Forensic science breaks down into the categories of:

- Forensic Science.
- Forensic Medicine.

- Forensic Psychology.
- Forensic Dentistry.
- Forensic Graphology.
- Forensic Accounting.

Each of these categories will at some time or other be employed in a criminal investigation and can also be used in commercial matters such as insurance investigations, allegations of fraud and/or corruption, and proving the paternity of an individual in an ongoing investigation relating to child welfare.

FORENSIC SCIENCE AND CRIME

All of these categories have their own particular specialities and it is important to recognise that the role of Forensic Science in the area of criminal prosecution has proved vital in bringing many criminals to justice who might have otherwise slipped through the net. It is important to remember also that Forensic Science is not simply investigating the events surrounding and leading up to death. It is also used to prove that serious sexual and grievous bodily assaults have taken place and it has proved fundamental in how the legal system conducts cases once they have been brought before a court of law.

FORENSIC SCIENCE DEVELOPMENTS

In the last two decades forensic science has seen many important discoveries made and has had many different procedures added to its remit in order to aid in the detection of criminals and also in proving the innocence of people suspected of crimes they did not commit.

Indeed without forensic science it is fair to say that many of the crimes that are committed today in our world would go unsolved and most definitely unpunished if it were not for these advances. Forensic science has now become a scientific study in its own right and now includes some of the world's most important discoveries as part of its remit.

These include:

- DNA coding.
- Toxicology: The study of drugs and their effects on the human body.
- Serology: The study of bodily fluids.
- Ballistics: The study of firearms.

In addition to these studies forensic science also looks at time of death, cause of death and how a body is disposed of. It also takes into account the fact that bodies many not necessarily have died where they are located and techniques within the field can provide definitive proof of this through soil analysis, examination and categorising of various pollens, and many other factors such as temperature and weather conditions. It is fair to say that forensic science has become one of the most important areas of scientific study of the 20th and 21st century and will remain so for a long time to come.

FORENSIC SCIENCE'S BLOOD AND DNA TEST METHODS

Detectives once relied on well-honed instincts and a suspect's confession as the best methods of solving a crime. In the late 1800s, fingerprinting changed all that; however, it was the advent of forensic science's blood and DNA analysis techniques that transformed police work into a true science.

EARLY METHOD OF DETECTING HUMAN BLOOD EVIDENCE

In 1900, a series of experiments by Paul Uhlenhuth led to a serum that reacted to the presence of human blood, but not to the blood of animals. The crime investigator would dissolve blood evidence in salt water, and then add the solution to the special serum. Human blood darkened the serum via the precipitation of a protein. The 1904 slaying of nine-year-old Lucie Berlin was the first murder case solved by the amasing new blood detection method.

CATCHING CRIMINALS WITH BLOOD TYPING

The first blood transfusions often resulted in the patient's death. Over time, scientists and physicians figured out that blood comes in different types, including A, B, O, and the AB "type-lacking" group. This discovery was immeasurably helpful in medicine, but had profound effects on the field of forensic science as well.

Blood groups were used to help eliminate certain suspects and allow the police to zero in on others. Still, such tests had serious limitations: Certain blood types could not be identified in the presence of other blood types. When a crime scene presented bloodstains that were a combination of several different types of blood, the evidence's value often plummeted. In 1966, English biologists Margaret Pereira and Brian Culliford discovered that a person's blood protein was as individual to her as a fingerprint. This discovery was considered the last word on blood evidence techniques until "DNA fingerprinting" eclipsed it.

DNA EVIDENCE: SOLVING CRIMES WITH THE GENETIC CODE

Deoxyribonucleic acid is present in every human cell; save those without nuclei, such as red blood cells. Its double helix structure was discovered in 1933 by Francis Crick and James Watson. Analysis of a person's DNA reveals how his genetic code resembles that of others and how it differs from theirs, too. Genetic sites (also called polymorphic loci) that differ the most from person to person are the most useful in identifying a suspect as the perpetrator of a particular criminal act.

In 1980, Sir Alec Jeffreys developed the original DNA fingerprinting method by studying repeating sections of genetic material called VNTRs (variable number tandem repeats), which make up the non-coding or "junk" sections of the DNA molecule. From this early work and its subsequent refinements, crime scene analysis has been positively revolutionised. After the laboratory testing procedures are complete, a specific formula is used to calculate the probability of a randomly selected person having all the same DNA markers as the DNA collected at a crime scene.

DNA TESTING HAS NEGATIVE AND POSITIVE ASPECTS

Sex offences and murders usually involve DNA evidence, which can lead a literal trail right to a suspect's doorstep. DNA results have also been used to free an unjustly convicted defendant. Of course, no scientific test is perfect. Identical twins cannot be differentiated using DNA analysis. Extreme care must be taken to prevent sample degradation, contamination, or cross-contamination between various samples. Human error, bad laboratory practices, or even outright fraud can still lead to skewed results. Future improvements in testing techniques will continue to further the accuracy and integrity of DNA as a true 21st century crime solver.

CODES OF ETHICS IN FORENSIC SCIENCE PRACTICE

Organisations like the American Academy of Forensic Sciences have developed a Code of Ethics and Conduct that the organisation enforces. The Ethics Code states that members 'shall refrain from providing any material misrepresentation of education, training, experience or area of expertise'. It also requires members to 'refrain from providing any misrepresentation of data upon which an expert opinion or conclusion is based'. Additional provisions are part of the Code of Conduct. They are designed to prevent members from falsely claiming to represent an organisation and from engaging in conduct adverse to the best interests and purposes of the organisation. Such transgressions of a Code of Conduct may violate the rules of the organisation but may not represent violations of fundamental forensic scientific ethics. The Code of Ethics and Conduct also has provisions describing due process procedures.

When enforcing a Code of Ethics, the following questions, as enumerated by Rosner, should be answered:

- What provision is the member accused of violating?
- What are the criteria for that provision?
- What are the relevant data?
- What is the reasoning by which it is determined that the member has or has not violated the specific provision?

It is important to write ethics provisions clearly in order to prevent ambiguity that could result in an ethics hearing degenerating into a popularity contest. For example, in the absence of clear guidelines, there is a danger that a verdict could primarily become dependent on whether a hearing panel likes an accused individual, whether the accused person's views on a controversial issue are popular, or whether the accused is part of the 'in' group. As a result, the actual seriousness of the ethics offence would become much less relevant, with less effort expended towards determining exactly what happened and for what reason, thus obscuring the stated goal of the hearing. If there are sections of the Code that potentially are unclear to a reader, clarifying interpretations should be disseminated to the organisation's membership.

It is essential for members to have information in advance about the specific types of conduct that are prohibited (minimal standards), with a clear distinction from those that are encouraged (aspirational standards). If there is any potential ambiguity, a way must be found to provide the necessary clarification. One possible means is to print an explanation of the issue in the organisation's newsletter.

STANDARDS FOR GOOD FORENSIC PRACTICE

Aspirational standards are standards of good practice or even excellence. Good forensic practitioners should strive to attain these standards despite the fact that only minimal basic ethical standards are subject to enforcement.

The American Academy of Forensic Sciences' Committee on Good Forensic Practice has developed the following Standards for Good Forensic Practice:

- Forensic scientists generally should follow the standards of their respective disciplines. They should apply with care any assessment methods, technical skills, scientific and other areas of specialised knowledge, to legal issues and questions. They should always strive to do high quality work.
- Forensic scientists should strive to keep current and maintain competence in their scientific disciplines. Although competence at minimum should be a goal, forensic scientists should strive for excellence.
- Forensic scientists should demonstrate honesty and should strive for objectivity, by examining scientific questions from all reasonable perspectives and by actively seeking all relevant obtainable data that could distinguish between plausible alternative possibilities.
- Forensic scientists should strive to be free from any conflicts of interest. They should possess an independence that would protect their objectivity. Any potential conflicts of interest should be disclosed. Work on relevant cases should be avoided or discontinued if objectivity may be compromised.
- Forensic scientists should undertake cases and give opinions only in their areas of expertise, attained through education, training and experience.
- Forensic scientists should attempt to identify, deter and help eliminate unethical behaviour by other forensic scientists through methods such as discussion with a colleague, education, and if unsuccessful, by filing an ethics complaint.

- It is essential to recognise that honest differences of opinion exist and do not imply unethical behaviour by either expert. The legal adversary system includes opposing attorneys seeking out experts with favourable opinions. Forensic scientists should not be blamed unfairly for unpopular verdicts, honest differences of opinion, or the vagaries of the legal system.
- Passions against an opposing disagreeing expert, or personal animosity, should not constitute the basis for an ethics complaint. Ethics complaints must be made in good faith. If based on passion alone, such ethics complaints themselves are inappropriate.
- Forensic scientists should present their opinions to the trier of fact in concise understandable language, but care must be taken since such efforts can result in oversimplification and loss of some precision. In their efforts to communicate effectively, forensic scientists should strive to be as accurate as possible and avoid distortion. Every reasonable effort should be made to ensure that others (including attorneys) do not distort the forensic scientist's opinions.
- Forensic scientists should strive to instill the highest ethical and scientific standards in their students and colleagues through such means as teaching, supervision, setting a good example, publications, and presentations at meetings.
- Forensic scientists should strive for excellence and the highest degree of integrity. Forensic opinions should not be based on undisciplined bias, personal advantage, or a desire to please an employer or an attorney.
- When forensic scientists are asked to express opinions on a legal issue, they should make every effort to become familiar with the applicable legal criteria in the pertinent jurisdiction. They should take care to reach only those legal conclusions that result from proper application of the data to that legal issue.
- Unlike attorneys, forensic scientists are not adversaries. They take an oath in court to tell the whole truth. They should make every effort to uphold that oath.
- When a forensic scientist accepts any privileged information from an attorney, care should be taken to ensure that all such information is kept confidential and does not reach the opposing side. After accepting such information, forensic scientists should not provide their services to the opposing side unless legally ordered to do so. Forensic scientists should alert attorneys not to make payment or provide privileged information, if they wish to retain the option to be employed by the opposing side.

ETHICAL PROBLEMS IN THE FORENSIC SCIENCES

Some forensic scientists confuse their role with that of an attorney. The attorney's role is to present the best one-sided case for his/her client, the only exception being not to argue for anything the attorney knows to be untrue. Unlike an expert witness, attorneys take no oath in court. In contrast, forensic scientists take an oath 'to tell the truth, the whole truth, and nothing but the truth'. If forensic scientists assume the attorney's total advocacy role and present the best one-sided case for the retaining attorney regardless of the 'truth' and the oath taken, the forensic experts may be perceived as 'hired guns'. They thereby vilify not only their own reputation, but also taint that of the entire field. Many forensic scientists consider the 'hired gun' problem the most serious ethical problem in forensic science.

A 'hired gun' can establish a track record of testifying for both the plaintiff (prosecution) and defence, depending on the side that hires the expert. An effective 'hired gun' can appear impartial and objective by giving one-sided misleading persuasive explanations for whichever side hires him/her for a particular case. 'Hired guns' not only always make the best case for the side hiring them, but alternatively they also could do so for a side for which they have a personal bias, regardless of which side they actually believe is right.

In contrast, it is possible to have a bias and preferentially work for one side, yet refuse to take on a case unless the forensic scientist honestly believes an opinion is true with the greatest objectivity possible. The distinction between a hired gun and an expert with an idiosyncratic opinion is not always clear-cut without knowing what is in the mind of the expert. Even if experts firmly believe an opinion because of personal bias so that they honestly may be telling what they believe to be true, others frequently might incorrectly view the opinion as a dishonest distortion of the truth. Sometimes, the problem is an insufficient effort by the expert to be objective, or an honest difference of opinion between forensic scientists.

Honest differences of opinion do exist as they do in every science and profession including law, and it is unfair to blame forensic scientists and forensic science for the battle of the 'experts'. Additionally, once a forensic scientist reaches an honest objective opinion, the pressures of our adversary system encourage experts to advocate for their opinion. It is difficult to remain totally objective when defending an opinion. Subtle pressures can also bias an expert, such as the wish to please an attorney or an employer. An honest expert should make every effort to form opinions based on the evidence even if contrary to the wishes of the hiring person or institution.

There are also questions about what it means to tell the truth, the whole truth and nothing but the truth. Does it mean the expert should try to offer the whole truth and not a misleading part of the truth to the degree that the attorneys and the court will permit? Or is it enough to present only that part of the truth that will help the case and rely on good cross-examination (which may not happen especially if the case is settled before trial) to bring out the rest of the truth? Or should some portions of the truth that may lead to conclusions contrary to the expert's opinion be presented solely as a tactical maneuver to make a pre-emptive strike so that those aspects can be debunked before the other side has a chance to present them? Should a forensic scientist try to participate in the legal system only in ways that further justice or should the forensic scientist solely answer the questions asked and trust the adversarial system to usually achieve justice?

The adversarial approach exists in those countries like the United States that base their system on English common law. Should forensic scientists lend their expertise to sides trying to achieve an 'unjust' outcome? Is it presumptuous for forensic scientists to try to further only what they see as justice or what is consistent with their professional scientific or personal values? Answers to these questions presently lack a consensus and a resolution does not seem likely in the foreseeable future.

FOUNDATIONS OF ETHICS

Are ethical guidelines arbitrary rules, or are they based on something more basic and fundamental? Ethical codes address 'important' issues, and seem to represent more than codes of conduct or codes of etiquette. Sometimes codes confuse etiquette (such as not criticising another scientist or professional or not hurting an organisation) with fundamental ethical issues. There are questions whether any ethical truth is objectively right or wrong, or whether all ethics is subjective and one code is as good as another. Some, who view ethics as subjective, see it as solely dependent on the mores of a specific culture or scientific discipline. They believe that what is considered ethical is only what feels morally right to individuals in that culture or group.

Some consider that the only foundation for ethics and morals is religion. However, religion itself lacks objectivity not only within the confines of a single religion, but also because there would be no consensus as to which of many different religions should be the standard bearer. Some religions in the past have advocated human sacrifice.

Did that make it 'right' for them? There is a long tradition of secular ethics independent of religion in western democratic countries. Although religion can be a basis for ethics, many religiously observant people can be unethical and do 'wrong' things despite engaging in religious practice. Atheists can be very ethical and committed to philosophical study to determine what is 'right'. Although religion can undoubtedly motivate many people to do what is 'right', by no means is religion a necessary foundation for ethics and morals.

Two primary schools of thought exist regarding the basis for ethics. One is consequentialist and proposes that the ethical thing is whatever leads to the best consequences, such as the most good or happiness for the most people. A subset of this position is utilitarianism, insofar as what is most useful is ethical. Another school is deontological and bases ethics on an intrinsic duty. Actions are intrinsically right or wrong regardless of consequences.

An example of the latter is the philosopher Immanuel Kant's categorical imperative – to do what you believe any ethical person in your position should do, such that the maxim you are following should be a universal principle for all to follow. There can be problems with following any of these positions too rigidly.

You might consider justice an intrinsic deontological duty, much as Kant did. Does that mean though that every crime no matter how small should be investigated until the perpetrator is punished? It takes consequentialist considerations to balance this duty with the desirability of spending government money on other things that might be more important for society's welfare. Investigating every crime thoroughly until the criminal is caught would use up all our society's resources. No funds would be left for other desirable purposes. Additionally, if deontological considerations were primary, two deontological duties might conflict without any principle to resolve the conflict.

Similarly, following only consequentialist considerations results in problems. For example, it might decrease crime in society if it were perceived that every criminal was always swiftly caught and punished and no criminal ever got away with a crime. To achieve this end it might help to choose individuals from an unpopular group such as prior offenders, try them in secret, find them guilty, and execute them quickly and publicly. Such a procedure might in fact have some deterrent value and decrease crime especially if it were used for white-collar crimes like embezzling money or income tax fraud.

However, deontological considerations of justice convince us that there is something wrong with this procedure. There are even contrary consequentialist considerations. Government would become very powerful, and any of us could be arbitrarily tried, convicted, and executed even if we were innocent. All of us would be at potential risk. Thus, a purely consequentialist foundation for ethics would also provide no way to decide between these competing consequences to determine the right action.

The best approach requires a consideration of both types of ethical foundations, but there is no higher order rule to help us in balancing these considerations. The theoretical approach to resolving ethical dilemmas essentially uses universal ethical principles as 'axioms' from which particular ethical judgements are deduced as 'theorems'. A solution to a problem in a particular case is deduced from the ethical principle. However, there usually is no higher order ethical rule or principle to help us decide among competing ethical principles.

In contrast, the practical approach, known philosophically as 'casuistry', does not involve universal theoretical principles and rules but requires us to consider the issues from the perspective of a particular case. General ethical principles and rules essentially are 'maxims' that can be fully understood only in terms of paradigmatic cases that define their meaning and force. However, there can be and often are differences of opinion about which facets of a case should be overriding, with no ethical guidance to help us in weighing and balancing competing ethical maxims.

In both instances, deontological and consequential considerations can assist us in determining what is 'right'. However, when ethical rules and principles conflict, there generally is no rule, principle, or maxim to help us resolve the conflict. Therefore, it is essential that forensic scientists are able and prepared to think through and resolve ethical dilemmas themselves.

In difficult situations, more than one solution usually should be considered acceptable, and ethics committees should not place sanctions on forensic scientists for following either course. In order to help reach the 'best' solution in such cases, consultation should often be obtained from experienced forensic scientists knowledgeable about ethics issues.

ETHICAL DILEMMAS

Ethical dilemmas often defy consensus solutions. When ethical responsibilities conflict, some forensic scientists always give the legal needs priority. The majority, however, usually give the legal needs priority, but make exceptions if the law desires an action that violates serious aspects of the values and ethics of their own scientific discipline or profession. In such situations, their professional or scientific values, ethics, and responsibilities may outweigh the desires of the legal system, and preclude their involvement in that aspect of a legal case. Personal ethics and morals can also preclude such involvement. Although the legal system can establish legal ethics, only scientific, professional, and forensic science disciplines and organisations can establish forensic science ethics. The law cannot determine what is ethical in forensic science; it only can determine what is legal.

Ethical dilemmas occur when there are conflicting ethical duties. An example of such a conflict occurs in forensic psychiatry, in circumstances in which a practitioner might sometimes believe that the most accurate and truthful assessment would be attained by misleading the person being evaluated. Such deception is often legal, but it violates forensic psychiatric ethics as promulgated by the American Academy of Psychiatry and the Law. Most forensic psychiatrists follow the guideline of the forensic psychiatric profession regardless of the frequent legality of such deception.

In another instance, many consider the presentation of aggravating circumstances in a death penalty phase of a trial a violation of medical ethics, even if such aggravating circumstances are true and such testimony in the United States violates no current ethical requirement. The same is true if individuals incompetent to be executed are treated to make them competent. These problems occur when medical or psychological skills are used to evaluate an individual for forensic purposes.

The specter of ethical conflict, however, is by no means limited to forensic psychiatry and psychology or other branches of forensic medicine, but it occurs in all forensic sciences. The use of DNA evidence in the courtroom, although seemingly based on sophisticated modern scientific techniques, with unusually high levels of reliability and certainty, is another example of a highly controversial area. Problems in other aspects such as data collection can be questioned. In many ways it seems counterintuitive that there can be such divergence in professional opinions when accurate data such as DNA evidence are introduced.

Nonetheless, controversy does exist about the interpretation of DNA data among highly respected forensic scientists, such as in the widely publicised criminal trial of O.J. Simpson. Does that necessarily mean that one side is unethical or is it possible for honest forensic scientists to hold strong opposing views? One possible explanation is that the highly accurate nature of DNA evidence has made many problems more significant. Potential ethical and data collection problems that had been readily ignored when identification methods were less accurate suddenly have begun to loom in importance. Many such controversies exist in all forensic science disciplines.

PROBLEM OF RISK IN FORENSIC SCIENCE

The last decades have seen a notable increase in the use of scientific evidence in criminal investigation, a development which has captured the imagination of the media, publics and policymakers alike. However, questions also continue to be raised about the judicial failures which continue to arise due to the alleged faulty use of forensic science. Concerns have continued to be expressed over the assumptions, techniques and practices used by forensic scientists in a number of areas of work. Such concerns have contributed to a growing interest in the adoption of more risk-conscious epistemological approaches, in an attempt to render forensic science reasoning processes not only more commensurable with conventionally regarded standards of scientific inquiry, but also in ways which acknowledge the risks to justice from the erroneous interpretation of forensic evidence.

According to some scholars, a new paradigm has arisen which accepts the conditionality of forensic scientific claims and the epistemic risks associated with the interpretation of evidence (Saks and Kohler 2005). This 'paradigm shift' has involved a move away from categorical assumptions of uniqueness of identification evidence, to a position which accepts epistemic risk in evidential interpretation, and advocates the use of probabilistic methods. Bayes Theorem has become prominent as a philosophical basis for the development of interpretation methods which reflect the conditions of uncertainty typical of criminal investigation.

The risk in forensic science, by way of a notable example. A focus on the Case Assessment and Interpretation (CAI) model demonstrates how forensic scientists have attempted to balance epistemic risk management concerns with the realities of the criminal justice system, which has itself been subject to a degree of neoliberal reform. We proceed to show how attempts to introduce this new sensibility have opened up areas of controversy regarding the role of forensic science vis-à-vis the wider criminal justice system within which it is embedded.

ORIGINS OF THE 'NEW PARADIGM' IN FORENSIC SCIENCE

Attempts to introduce Bayes Theorem into forensic science date from around the beginning of the 1960s, and can be traced back to the work of both forensic scientists and legal scholars. In what follows I briefly trace the origins of this development.

In the early 1960s, Paul Kirk, a leading figure in US forensic science, published an article entitled 'The Ontogeny of Criminalistics' in which Kirk reflected on the extent to which forensic inquiry could be considered a science in its own right. Marking a particularly notable attempt to delineate the disciplinary boundaries of forensic inquiry, Kirk defined criminalistics as the 'science of individualization' (Kirk 1963, p.236).

In a series of papers Kirk and his student Charles Kingston outlined the way in which an advanced appreciation of statistics and probability could be used to enhance 'interpretive areas of criminalistics', the first significant attempts to provide a systematic probabilistic approach to evidence interpretation (Kirk and Kingston 1964, p.514). In two papers concerning the 'Applications of Probability Theory to Criminalistics' (1965a, 1965b), Kingston discussed the use of probabilistic models for the assessment of certain forms of transfer evidence, and included a favourable consideration of an explicitly Bayesian model (Kirk 1965b).

Around the same time, legal scholars also began to consider how Bayesian probability could be used to overcome the risks of misinterpretation of evidence, as exemplified in controversial cases such as People vs Collins. Kaplan (1968) argued that decision theory, which had largely been developed in the fields of economics and business, and which utilised Bayesian probability, enabled a better understanding of the epistemological processes involved in the judicial arena (Kaplan 1968). In another influential article, Finkelstein and Fairley (1970) questioned the uniqueness principle which had predominated in forensic science.

They argued that it was both wrong and dangerous to attribute uniqueness to evidential data, and that in many cases, identificatory evidence cannot be claimed to be unique, merely rare (Finkelstein and Fairley 1970, p.516). In order to assess and express weight of evidence, they cited Bayes Theorem as an appropriate means for translating frequencies of rare events into a probability assessment. At the heart of their argument was the point that 'unique' characteristics could not be definitively attributed to an individual, but that certain infrequently occurring forms of transfer evidence could actually demonstrate significant probative weight if incorporated into a Bayesian formula. Hence they argued that individualisation could be quantitatively assessed, if not categorically proven, via the use of Bayesian reasoning.

The Bayesian approach to evidential reasoning began to receive further support, most notably via the influential work of Ian Evett, a statistician based at the UK Forensic Science Service (FSS). Evett collaborated with the distinguished Bayesian scholar Denis Lindley (Evett and Joyce 2005), and made significant advances in applying Bayesian approaches to trace evidence analyses. These and other studies significantly contributed to a growing perception of the potential of Bayes Theorem to provide a more scientifically robust basis for the assertions of forensic scientists. Evett saw the use of Bayes as an ideal solution to the unique problems experienced in forensic scientific work: That framework—call it Bayesian, call it logical—is just so perfect for forensic science. All the statisticians I know who have come into this field, and have looked at the problem of interpreting evidence within the context of the criminal trial, have come to see it as centring around Bayes's Theorem.'

The use of Bayes was seen as aiding the transition from a categorical epistemological position (the uniqueness assumption), to a conditional paradigm which aspired to improved standards of scientific propriety. In enabling the reasoner to update one's beliefs in a sequential, iterative manner, Bayes Theorem became viewed as being well suited to the contingencies of a criminal investigation. The adoption of a quantitative measure of subjective belief was also understood to provide a more accountable form of reasoning than the opaque, intuitive methods that had traditionally been associated with several forensic disciplines.

Bayesian reasoning now forms the basis of a wide range of forensic technologies and initiatives. A particularly notable development in this regard concerns the Case Assessment and Interpretation (CAI) model, which represents an attempt to construct an all-encompassing method of evidential interpretation for use in the course of criminal investigations. Developed by a team of FSS scientists, the CAI seeks to continue to shape the process of evidential reasoning in a way that logically accounts for epistemic risk, and hence promotes the new paradigm.

Furthermore, the CAI also reflects new commercial realities. UK forensic science has been subject to an increasing degree of commercialisation in recent decades. In 1991, in response to government reviews which urged the reform of forensic scientific support in line with the 'New Public Management' ethos, the FSS introduced a system of direct charging to forces and other clients (McFarland 2003). Forensic science providers such as the FSS hence entered into a new relationship with police, in which the latter became customers. This development occurred at a time when responsibility for force budgets became increasingly devolved. Hence the CAI was intended to reflect this new orientation, 'a radical change of culture' where the costs of forensic science were 'no longer invisible to operational policemen' (Cook et al 1998a, p.151). The aim of this initiative is therefore to provide better 'value for money' to customers, and to also give the customer greater say 'about what work is done in the laboratory' (Cook et al 1998a, p.151). At the same time, it is intended to increase the contribution of forensic scientists to criminal investigations, as pointed out by one of the authors of the CAI:

'There should be a focused, directed use of resources to help make progress with the matters that are of relevance and importance to the customer and for which forensic science can actually provide useful information. Wherever possible, scientists should be addressing questions that are a stage advanced beyond the simply analytical.' Hence the CAI is also intended to promote a more expansive form of scientific support.

Producing and Delivering Bayesian Forensic Inquiry

In broad terms, the CAI is delivered to police users in three overarching phases: 'customer requirement', actual 'case assessment', and 'service delivery' of written statements of test results together with an assessment of their significance. This process is sensitive to feedback, and continually subject to review with the customer; new lines of inquiry generated by subsequent developments can be readily incorporated into the framework.

The customer requirement phase involves a rigorous consideration of the client's specific needs. Here, the construction of an appropriate 'framework of circumstances' is taken to be a necessary pre-requisite for the development of propositions relevant to the evidence and the case. The authors stress the need for scientists to take a 'balanced view' of each case, in line with what they regard as the principles of 'the Bayesian view of evidence, that it is not sensible for a scientist to attempt to concentrate on the validity of a particular proposition without considering at least one alternative'.

In the case assessment phase, the scientist is expected to further clarify the propositions they wish to investigate by re-describing them in more quantitative terms, via the generation of likelihood ratios, through which the scientist seeks to organise the logical processing of evidential propositions using Bayes Theorem. These take the form:

- Probability of the evidence if prosecution proposition is true
- Probability of the evidence if defence proposition is true

The likelihood ratio (LR) enables the scientist to represent his/her expectations in a more precise and quantifiable manner. The LR also functions as a key device for enabling providers and customers to reach agreement on the most efficient course of action to take in the pursuit of an investigation. Cook et al (1998a) describe how the LR approach can be used to assess the usefulness of scientific tests in relation to specific cases. In one example they describe how this framework can be used to assess the significance of finding certain quantities of glass on the garment of an individual implicated in an alleged break-in, based on estimated probability distributions. Using these distributions, pairs of propositions, relating to the case, can be assessed, for example

- 'The suspect is the person who committed the burglary'
- 'The suspect has had no connection with the crime scene'

The probability estimates for finding certain amounts of glass may affect the value of the LR returned. In doing so, Bayesian reasoning provides a means of assessing the potential probative value of conducting glass analysis.

Table. Probabilities of Finding Quantities of Matching Glass

Quantity (Q)	Probability Pr (Q|a)	Probability Pr (Q|b)
None	0.05	0.95
Few	0.30	0.04
Many	0.65	0.01

Table. Likelihood Ratios for the Three Values of Q

Quantity (Q)	Likelihood Ratio (LR)
None	0.053
Few	7.5
Many	65.0

Table. Probabilities of Possible Outcomes From An Examination for Matching Glass

Opinion Expressed	Probability of Opinion if a is True	Probability of Opinion if b is True
Moderate Support for b	0.05	0.95
Weak Support for a	0.30	0.04
Moderate Support for a	0.65	0.01

From the information in Table above, it is possible to obtain LR values by dividing the proposition pairs corresponding to each matching fragment group, as reproduced in Table above. These can be further summarised in terms of the support they show for each of the propositions. This leads to the assessment that there is a 65% chance that the examination will provide moderate support for prosecution proposition 1a, and 30% chance that it will only provide weak support. If the suspect is innocent however, then there is a 95% chance of moderate support for innocence, although there is also 5% chance that the evidence will falsely incriminate him.

The Hierarchy of Propositions

In the above example, the analytical process is wholly the preserve of the scientist. Although the decision to conduct glass analysis would remain with the customer, the circumstances of this specific case do not enter into the consideration of the evidence. As indicated by the concern to 'add value' to investigations however (Jackson 2000, p.84), the CAI seeks to link these kinds of analysis with the wider framework of circumstances of a case, so that the investigative process can, in theory at least, continue to proceed along Bayesian lines. A key element of the CAI is the hierarchy of propositions, which plays a central role in organising the way in which propositions are constructed and assessed. This hierarchy classifies propositions in order of their relevance to the ultimate issue under consideration by the court (Cook et al. 1998b).

Level I, or the source level, relates to propositions concerning the origins of the evidentiary material, and.will exclusively involve the measurement and comparison of quantitative data (Cook et al 1998b, p.232-233). Level II in the hierarchy, the activity level, involves a greater element of reconstruction of the events in each case. Re-visiting the above example, one may wish to pose a prosecution hypothesis, concerning the probability that a given suspect smashed the window, against a defence hypothesis seeking to ascertain the probability that the suspect was not present when the window was broken. If the investigators wish to address the probability of finding a matching quantity of fragments on a suspect given that it was they who had smashed the window, they would require further circumstantial information.

In this case it would be of use to be able to ascertain how the window was smashed, and the time interval between the incident and the removal of the clothing from the suspect for analysis. In order to assess a corresponding defence hypothesis, in this case the probability of finding matching glass on the suspect if they were unconnected with the smashing of the window, information would be sought which would predispose the suspect to have glass present on his clothing, such as lifestyle or profession (Cook et al 1998b, p.233). Hence the more information a scientist may possess in relation to the circumstances of the case, the more fully Level II propositions will be able to considered, and the more bearing that insights generated at Level I will have on consideration of Level II propositions.

The emphasis on the role of forensic scientist in generating Level II propositions raises some questions about the division of ratiocinative labour in the course of investigating a criminal case. The kind of level II propositions outlined in the literature appear to be the kind of lines of enquiry that one would expect the police to pursue. Hence the remit of what could be viewed as 'forensic science' is possibly stretched; what the CAI can be seen to represent in this regard is a possible extension of what may be seen as constituting forensic scientific inquiry. Rather than a concern with the recovery, rendition and analysis of items of possible evidential interest and importance, 'forensic science' under the realm of the CAI, is extended to constitute a possible method for the advancement of criminal investigation as a whole.

The final level in the hierarchy, Level III or the offence level, concerns the probability that a suspect has committed a criminal offence. Level III propositions are the domain of the jury, assisted by the judge (Cook et al 1998b, p.233). Offence-level propositions may also be construed as activity-type propositions; however the key difference is that Level III propositions concern the question of whether an actual crime has occurred. Although Level III propositions are assessed by the jury, and ultimately outside the domain of the forensic scientist, the latter may well be able to assist the court in their deliberations, by reporting information concerning likelihoods in a suitably comprehendable form (Cook et al 1998b, p.233).

Contestations

The previous section has sought to indicate how the CAI has been specifically designed to embed Bayesian reasoning within the wider context of policing. As I describe below however, attempts to apply CAI have opened up new spaces of contestation between and within elements of the criminal justice system, including forensic science itself.

Within the framework of the CAI, the scientists are meant to act in a facilitative role, helping police to clarify the important questions to ask in the course of a case. For the CAI to function effectively, and for evidence to exert optimal impact on the course of a case if considered within this Bayesian approach, forensic scientists need potentially high levels of information from the police. The CAI also emphasises the desire for scientists to enjoy a greater degree of influence with regard to which pieces of evidence are seen as most pertinent to a case.

However, it became clear through field research that this potential re-positioning of forensic scientists within the decision-making hierarchy of investigative frameworks was not necessarily in keeping with traditional police views. Instead, it appeared that police investigators viewed the role of forensic scientists as somewhat more limited: to simply provide scientific data about evidence, which may occur in relative isolation to the deliberations made by police concerning the progress of a criminal case. Moreover, there appears to have been suspicion on the part of the police with regard to the need to consider defence propositions within the framework of the CAI:

- 'We had another group of officers who kind of mis-interpreted really...we had to re-assure them we weren't out to help the defendant...' (Interview, 2007)

Tensions also became apparent with another set of actors – the FSS management themselves. Although the CAI emphasised engagement with customers as facilitating an efficient use of the latter's resources, it appears that the CAI clashed with management interest in maximising revenue from the same set of actors:

- '...I think [the FSS management] felt some commercial problems... the way we were working in terms of earning our incomes was items examined, the more items we examined the more income we got. So there was almost a counter-pressure not to apply CAI, because

CAI in some ways, said 'lets just look at the items that are gonna be really effective, really efficient, in addressing this question, and if you decide with the customer these are the key issues in the case, the strategy to address these key issues is this, this and this, it's not these items or these tests... and I think therein lay some of the difficulties from the managers and leaders, because you could see the natural consequences, if we apply CAI…we're gonna lose a lot of income, potentially.' (Interview, 2007)

It was clear that the FSS management saw commercialisation in a different way. Rather than providing tests in a selective and targeted manner specific to customer needs, FSS management viewed forensic scientific 'products' as being provided to customers on a wholesale basis. In this view, a 'product' was packaged as a specific type of test, (e.g. DNA analysis), to be sold to customers in multiple 'units'.

- 'In the FSS now, they've gone away from experienced case scientists looking at multiple evidence types and synthesising all of the evidence together to neat little boxes of 'products' where 'you want the hair analysed?' (Interview, 2008)

Here it is possible to identify different ideas on what a 'commercialised' science might entail. The CAI developer's interest in maximising the customer's own value for money, runs contrary to the need for the business leaders to maximise readily measurable profit.

The CAI was also found to have created controversy within the forensic science community itself. Discussions highlighted a certain reluctance on the part of many forensic scientists to open up their methods to the kind of epistemological scrutiny which Bayesianism demands. Some practitioners appeared to be reluctant to do so:

- 'Another area for objection might be because we were challenging conventional wisdom, and we were almost challenging the basis of their expertise.' (Interview 2007)

Other forensic scientists disagreed with the manner in which Bayesianism had been adopted as a reasoning system. For example, one individual criticised the developers for conflating an adjunctive technique with the scientific method itself:

- 'Statistics is a tool. Its not science' (Interview, 2008)

2

Forensic Facial Reconstruction

Forensic facial reconstruction (or forensic facial approximation) is the process of recreating the face of an individual (whose identity is often not known) from their skeletal remains through an amalgamation of artistry, forensic science, anthropology,osteology, and anatomy. It is easily the most subjective—as well as one of the most controversial—techniques in the field offorensic anthropology. Despite this controversy, facial reconstruction has proved successful frequently enough that research and methodological developments continue to be advanced. In addition to remains involved in criminal investigations, facial reconstructions are created for remains believed to be of historical value and for remains of prehistoric hominids and humans.

DAUBERT STANDARD

In the U.S., the Daubert Standard is a legal precedent set in 1993 by the Supreme Court regarding the admissibility of expert witness testimony during legal proceedings. This standard was set in place to ensure that expert witness testimony is based upon sufficient facts or data, is the product of reliable principles and/or methods (including peer review), as well as ensuring that the witness has applied the principles and methods reliably to the facts of the case.

When multiple forensic artists produce approximations for the same set of skeletal remains, no two reconstructions are ever the same and the data from which approximations are created are largely incomplete. Because of this, forensic facial reconstruction does not uphold the Daubert Standard, is not included as one of the legally recognized techniques for positive identification, and is not admissible as expert witness testimony. Currently, reconstructions are only produced to aid the process of positive identification in conjunction with verified methods.

SKELETAL REMAINS IDENTIFICATION BY FACIAL RECONSTRUCTION

Facial reconstruction is a method used in forensic anthropology to aid in the identification of skeletal remains. The reproduction of the facial features of an individual is based upon the average soft tissue thicknesses over various anatomical sites of the skull and jaws (Krogman and Iscan 1986) and is duplicated by means of modeling clay. There are significant differences in the thicknesses of the soft tissues of males and females of different races. Various techniques have been employed to measure the thickness of the facial tissues of adults, children, and young adults. The methods used to "flesh out" a face may vary, but each method incorporates a harmonious balance between science and art that eventually results in a reproduction of a face that may lead to an identification.

In the Western Cape Province of South Africa, forensic pathologists examine numerous cases involving skeletal remains each year (Schwar et al. 1987). Most of these cases are the result of unnatural deaths and require forensic investigation. The usual procedure is to analyze the bones to determine the age, race, and sex of the individual (De Villiers 1968) and to correlate this information with the missing persons list. In most cases, attempts to identify the bodies of unknown individuals are limited to these examinations due to the high death rate in South Africa.

The local police are inundated with investigations of unnatural deaths, and most of the unclaimed skeletal remains are buried in numbered graves. When the case warrants further investigation, the author has undertaken facial reconstruction of a skull. The method used in these cases is that advocated by Neave (1979). The purpose of this paper is to report three cases of positive identification through facial reconstruction of the skeletal remains of six unnatural deaths that occurred in the Cape Province in South Africa.

Case 1

In 1994 a Dutch tourist who wandered off the usual pathway found the skeleton of a young Caucasian female on the summit of Table Mountain (Phillips et al. 1996). The remains were lying in an area of thick reeds that was approximately 200 meters from the pathway. Several personal items located in the vicinity of the skeleton were recovered, including a wristwatch, checkbook, shark-tooth pendant, and a leather handbag containing a plastic water bottle, a cup, and two empty packets of sleeping pills containing diphenhydramine. The number of tablets missing from the blister-packs would have amounted to a total toxic dose of 20 to 24 milligrams per kilogram of body mass (fatal dose = 25 mg/kg). The winter rainfall in Cape Town and the decomposition of the reeds led to the total obliteration of the printing in the checkbook, and there were no other identifiable items. The skeletal remains were collected and transported to the cable car station.

While awaiting the arrival of the cable car, a newspaper reporter overheard the conversation between the forensic pathologist and a police officer regarding the skeleton. She had covered the story of the disappearance of a young woman on Table Mountain six years previously. The young woman who disappeared had been suffering from chronic depression and was being treated by a psychiatrist. An extensive search was launched after her disappearance by the police, the army, and the airforce but was unsuccessful in locating the body. The location of the body in the tall reeds away from the usual pathway was one of the factors that contributed to its lack of discovery. The other factor was that most people who disappear on the mountain fall off the slopes, and the bodies are easily spotted from a helicopter.

The information from the reporter led the investigating officer to the parents of a missing 26-year-old woman. When the parents were shown the handbag, wristwatch, and shark-tooth pendant, they were relatively certain that these belonged to their daughter. Unfortunately, the dental records of the missing woman had been mislaid since the initial investigation six years previously, and this precluded positive identification. The skull and mandible were referred to the author for facial reconstruction as a final attempt at identification. The author and the artist had not seen any photographs of the deceased, and this afforded them the opportunity to construct a face on the skull and then compare it with known photographs.

The parents of the deceased were invited to view the facial reconstruction, and the comments of the mother are as follows: "Although the sculpture does not look exactly like our daughter, the family resemblance is remarkable, so much so that it looks exactly like our niece. We are satisfied now that our daughter is dead."

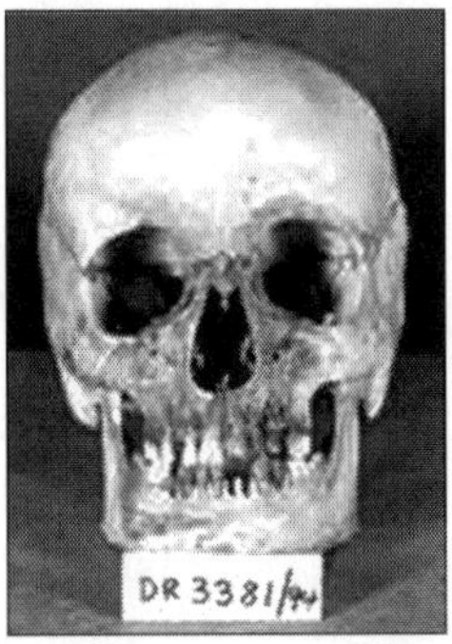

Fig. The skull found on the summit of Table Mountain among the reeds. The brown teeth are a result of staining from the organic compost.

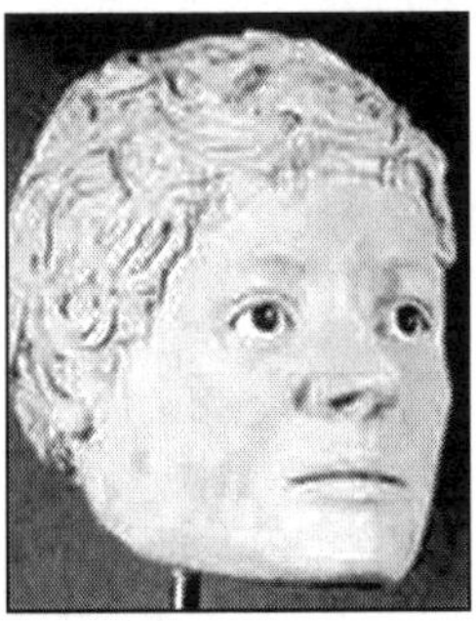

Fig. The sculptured face in modeling clay on a plaster model of the skull.

Fig. A photograph supplied by the parents at the time of the disappearance of their daughter.

Case 2

During the 1990s, an era of severe political unrest in South Africa, numerous assassinations took place throughout the country. Political rivals were eliminated, and the bodies were buried in remote locations only to be discovered by chance, in some cases, when excavations were undertaken for building purposes. The skeletal remains of four humans were discovered in 1995 buried in the sand dunes adjacent to the mouth of the Imzimvubu River in the Transkei region of the Eastern Cape, near Port St. Johns. The skeletal remains were submitted to the Forensic Laboratory at Salt River in Cape Town for analysis that included examination by a physical anthropologist.

Two of the skulls were aged at 20 to 30 years, one at 18 to 20 years, and the fourth at 14 to 16 years. After metric analysis of the skulls (De Villiers 1968), all were considered to be predominantly of Negroid origin. One of the older skulls had a bullet hole in the mastoid area that indicated these young people could have been victims of political violence. Using the average soft tissue thicknesses for the Negroid as published by Rhine and Campbell (1980), the faces of the four victims were reconstructed in clay over plaster models of the skulls. The mandible of the youngest victim was not submitted with the skull and was constructed before the facial reconstruction process.

Photographs of the facial reconstructions were shown to the relatives of the victims. The parents of the four missing youths then submitted photographs of their children for comparison. The parents of the victims were satisfied that their children had been identified.

Case 3

In April 1995, skeletal remains were unearthed in Crawford, Cape Town. The skeleton lay in a shallow grave and was entangled in female clothing. The skull was analyzed metrically to determine the age, race, and sex of the victim. The anatomical features of the skull were determined to be of mixed racial origin, containing Khoisanoid, Negroid, and Caucasoid features.

Gustafson's method of age determination of the teeth (Gustafson 1966) was performed, and the age was estimated to be approximately 24 years. The reconstruction of the face was carried out on a plaster model of the skull in the manner advocated by Neave (1979) and using the soft tissue thicknesses for persons of mixed racial origin as published by Phillips and Smuts (1996).

Following completion of the reconstruction of the face, it was photographed, and the photograph was placed in the newspaper for possible recognition. This was unsuccessful, and a further attempt using media exposure showed the photograph on a television program Crime Stop. This led to a telephone call from a woman living in Upington, Northern Cape, whose daughter had disappeared after attending a birthday party in Cape Town. A photograph of her daughter was subsequently submitted, and the mother was confident that her daughter had been identified.

Discussion

The sculpting of a face on a skull can be traced to biblical times. Forensic scientists and physical anthropologists revived this technique as a means of identifying human remains. Several forensic scientists have criticized this method of identifying skeletal remains, citing the lack of scientific reproduction of the final product and the low statistical success rates. However, the facial features that are achieved through reconstruction are not expected to be an exact replica of the person to be identified. The success of any facial reconstruction is the reproduction of facial features on a skull that may lead someone to suggest that the face reminds him or her of a particular person who is missing.

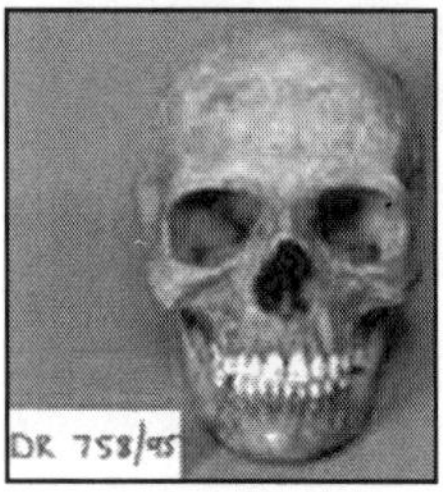

Fig. The skull (DR 758/95) of a young woman whose skeletal remains were found in a shallow grave in Crawford, Cape Town.

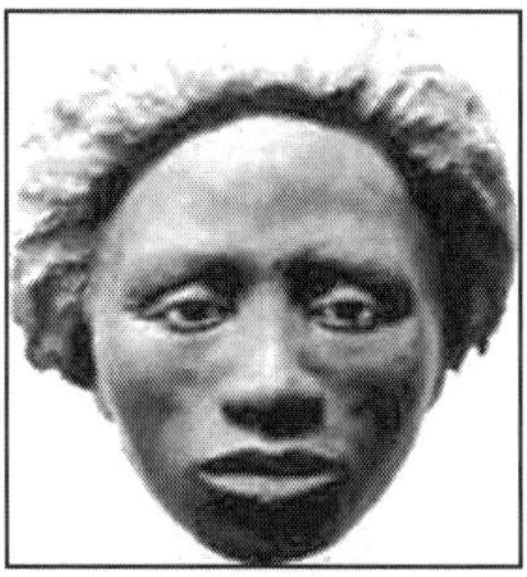

Fig. The reconstructed facial features on a plaster model of the skull (DR 758/95).

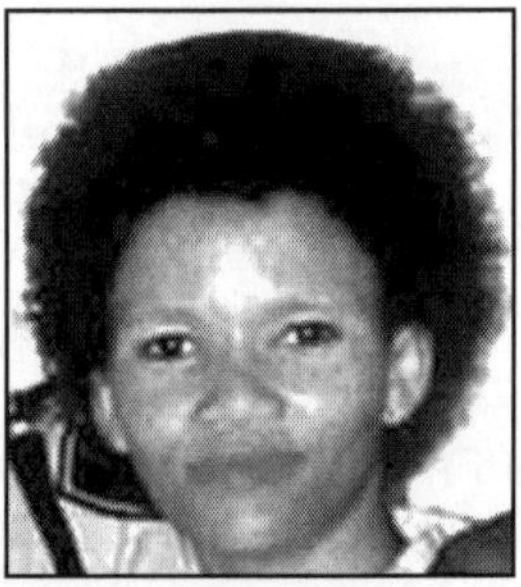

Fig. A comparative photograph sent by her mother.

TYPES OF IDENTIFICATION

There are two types of identification in forensic anthropology: circumstantial and positive:

- Circumstantial identification is established when an individual fits the biological profile of a set of skeletal remains. This type of identification does not prove or verify identity because any number of individuals may fit the same biological description.
- Positive identification, one of the foremost goals of forensic science, is established when a unique set of biological characteristics of an individual are matched with a set of skeletal remains. This type of identification requires the skeletal remains to correspond with medical or dental records, unique ante mortem wounds or pathologies,DNA analysis, and still other means.

Facial reconstruction presents investigators and family members involved in criminal cases concerning unidentified remains with a unique alternative when all other identification techniques have failed. Facial approximations often provide the stimuli that eventually lead to the positive identification of remains.

HOW FORENSIC FACIAL RECONSTRUCTION IDENTIFIES SKELETONS

In today's media, with numerous crime shows featuring the latest, cutting-edged technology, forensic facial recognition has reached the mainstream populace. However, in reality, forensic facial reconstruction is, for many reasons outlined below, typically used only as a last (rather than first) resort. Forensic facial reconstruction, also known as forensic facial approximation, is a technique in the field of forensic anthropology. This technique originated in two of the four major subfields of anthropology: biological anthropology (to approximate the appearance of early hominids) and archaeology (to visualize the remains of historic figures). In this technique of facial reconstruction, the skeletal remains – including the facial bones of the skull – are used to identify such characteristics as age, race, and gender in order to reconstruct the facial tissue and corresponding features. The goal of this process is to either discover the identity of the deceased or to reconstruct the features of a known decedent; examples include murder victims, historical figures, and prehistoric humans and hominids.

Forensic facial reconstruction is viewed by many as a questionable technique. This is due to its undeniable shortcomings. First, the data used to determine facial tissue thickness is as of yet very limited in terms of ranges of ages, sexes, and body types and builds – a disparity that can significantly affect reconstruction accuracy (Rathbun, 1998).

Second, facial reconstruction has also faced both criticism and skepticism due both to its subjective nature and its methodological inconsistency – the result of a combination of anatomy, osteology, forensic science, anthropology, and artistic sculpture. In fact, when numerous forensic reconstruction artists create approximations of a single set of skeletal remains, the reconstructions differ and the data are incomplete (Helmer et al.). This inconsistency is due to the artistic subjectivity of the facial reconstruction artists; however, it is worth noting that the shape and position of the main facial features can be fairly accurate, as such features are guided by the shape of the skull (Helmer et al.).

Third, as a result of this data insufficiency and subjectivity, a lack of reliability and standardization of a scientific method exists. As a result, facial reconstruction artists do not qualify as an expert witness under the Daubert standard federal rule of evidence and thus facial reconstruction artists can not testify as expert witnesses. Thus, they are only utilized in conjunction with verified methods in order to ascertain victim identification.

However, in spite of its challenges, the success and subsequent use of facial reconstruction has resulted in an advancement of both research and methodology (Forensic facial reconstruction, (Wikipedia). For facial approximation presents both investigators and family members with a creative identification tool when other such tools and techniques lack success. Additionally, facial approximations often provide the stimuli that eventually lead to the positive identification of remains.

History

Three-dimensional facial reconstructions as a scientific technique date back to the 1800s (Rhine, 1998). German anatomists Welcker in 1883, His in 1895, and, in 1898, Kollmann and Buchly, were the first to reproduce three-dimensional facial approximations from cranial remains through the use of tissue thickness measurements on identifiable points on the face and skull (Rhine, 1998). In a 1962 book, Wilton M. Krogman introduced his method for facial approximation (Iscan, 1993). Then, in 1964, Mikhail Gerasimov likely conducted the first attempt at the facial reconstruction of paleo-anthropological ancient peoples (Iscan, 1993). Later on, Gerasimov's students used his facial approximation techniques in criminal investigations. (Forensic facial reconstruction, Wikipedia.) Additional anatomists who helped popularize three-dimensional facial reconstruction include: Cherry (1977), Angel (1977), Gatliff (1984), Snow (1979), and Iscan (1986) (Reichs and Craig, 1998).

Identification Types

Two types of facial reconstruction identification exist: circumstantial and positive (Burns, 1999). Circumstantial identification takes place when a deceased individual's skeletal remains fit a biological profile (Forensic facial reconstruction, Wikipedia). Note that this identification does not identify the remains, but merely the biological profile. Positive identification – actual identification of an individual's remains – is the goal of forensic science. This precise identification takes place due to a specific and unique set of biological characteristics that apply to a given individual, such as medical records, dental records, DNA analysis, ante mortem wounds, and pathologies.

Reconstruction Types

The methods used to "flesh out" a face in the process of reconstruction may vary, but each one incorporates a creative balance between art and science that leads to a facial reproduction that could result in an identification (Phillips, 2000). There are three types of reconstructions: two-dimensional, three-dimensional, and superimposition.

Two-dimensional (2D) facial reconstructions are collaborations between artists and forensic anthropologists based on the deceased's skull and ante mortem photographs, with occasional (if less accurate) skull radiographs also used. Karen T. Taylor of Austin, Texas pioneered a commonly used method of 2D facial reconstruction during the 1980s. Taylor's method involves the placement and photography of tissue depth markers on an unidentified skull at various anthropological landmarks.

These photographic prints are then used to create facial drawings on transparent vellum. Today, F.A.C.E. and C.A.R.E.S. computer software programs produce, edit, and manipulate two-dimensional facial approximations. Although these computer programs offer speed and efficiency, their images are more generic than those provided by hand-drawn artwork (Reichs and Craig, 1998).

Taylor's 2D facial reconstruction process, including successful identification.

Three-dimensional (3D) facial reconstructions consist of two types: clay or plaster sculptures from casts of cranial remains and high-resolution, 3D computer images. Like 2D facial reconstructions, 3D facial reconstructions typically involve a partnership between an artist and a forensic anthropologist.

While sculptures are made from casts of cranial remains, 3D computerized facial reconstructions manipulate scanned photographs of the cranial remains, stock photographs of facial features, as well as any other reconstructions available. In victim identification, because they appear more realistic, computer approximations are usually more effective.

A 3D sculpture facial reconstruction comparison (left to right): the skull – labeled "DR 758/95" – from the skeletal remains of a young woman that were found in a shallow grave in Crawford, Cape Town; the plaster sculpture based on the skull; and a photograph sent by the victim's family. (SOURCE: Phillips, 2000)

Superimposition is a third technique of forensic facial reconstruction. Unlike 2D and 3D facial reconstructions – in which the identity of the skeletal remains are completely unknown – forensic superimposition requires knowledge on the part of investigators about the identity of a given set of skeletal remains.

In this method, a photograph of a given individual's identity that is suspected of belonging to the unidentified skeletal remains is superimposed over the unidentified skull's X-ray. If the skull and the photograph match up anatomically, then the skull and the photograph are of the exact same individual (Lundy, 1986).

ACCURACY OF FACIAL RECONSTRUCTION

The three-dimensional facial reconstruction technique discussed in this paper involves the production of facial sculptures onto the skull or skull replica. This approach involves modelling the facial musculature before applying a skin layer to depict the living facial appearance. Before dissecting the facial reconstruction process in detail the accuracy of this technique should be discussed.

The level of accuracy is of great importance to practitioners and law enforcement agencies. Without question, facial reconstruction has been a valuable tool for forensic investigation and many individuals have been successfully identified as a direct result of a publicity campaign employing a facial reconstruction. Practitioners report varying success rates, but it is unclear from the success rates to what extent the facial reconstructions were directly responsible for recognition and thus identification.

Typically, practitioners will show an image of the reconstruction next to an ante-mortem photograph of the identified individual to illustrate the accuracy of the technique. As only the successful cases are shown in this way, this is not an impartial assessment, and blind studies must be utilized to rigorously analyse the reliability of the techniques. However, the inherent flaw in the majority of blind studies is that it is practically and ethically difficult to represent a forensic scenario based on familiar face recognition (as access to skulls of known identity along with access to relatives of the deceased is almost impossible to achieve), so these blind studies often rely on unfamiliar face recognition and evaluation. The problems associated with the recognition of unfamiliar faces who recorded extremely high error rates in the verification of identity from photo-ID cards. Who investigated matching of unfamiliar target faces from high quality video stills against photographic arrays. The recognition rate was only 70% (where the rate by chance was 10%), despite the fact that the target still was taken on the same day as the array photograph. The recognition rates decreased further when unmatched views or expressions were employed.

A number of different methods have been employed to assess accuracy and reliability, including face pool assessment, resemblance ratings and morphometric comparison. The use of resemblance rating assessment has been criticized as misleading and insensitive, and face pool assessment and anthropometry are currently encouraged as more meaningful measures of the accuracy of facial reconstruction. Face pool assessment involves the comparison of an image of the reconstruction with a pool of face photographs that includes the target alongside a number of other faces of similar age, sex and ethnic group.

Volunteers are then asked to choose the face from the pool that most resembles the reconstruction, to determine the correct recognition rate. The level above that recorded by chance suggests the accuracy of the facial reconstruction. For the Manchester method analysed five female juvenile reconstructions using face pool assessment and recorded an overall recognition rate of 44% (where the level by chance was 10%).

A further blind study (Wilkinson et al. 2006) analysed two adult computerized reconstructions and recorded an overall recognition rate of 71% (where the level by chance was 20%). This study employed images of facial surface scan data in the face pool rather than photographs. These results compare favourably with previous reconstruction studies, where results range from 8% correct identification rate to 46.5%, and with the psychology study by Bruce et al. (1999), where recognition even from photographic in vivo images was only 70% (where the level by chance was 10%).

It may be possible to assess accuracy by morphometric comparison between the reconstruction and the target face. With the development of CT imaging and computerized facial reconstruction it is possible to compare the surfaces of the reconstruction and the target face using 3D modelling software (RAPIDFORM).

The Manchester method was evaluated in this way (Wilkinson et al. 2006) using CT data from two adults, one male and one female. The results suggested that 67% of the facial reconstruction surface had less than 2 mm error, even with the cheek distortion exhibited on the target face due to the position in the CT scanner. The study suggested that the nose, eyes, jaw line, forehead and chin could be reconstructed reliably (< 2 mm error), with the mouth and ears showing the most errors.

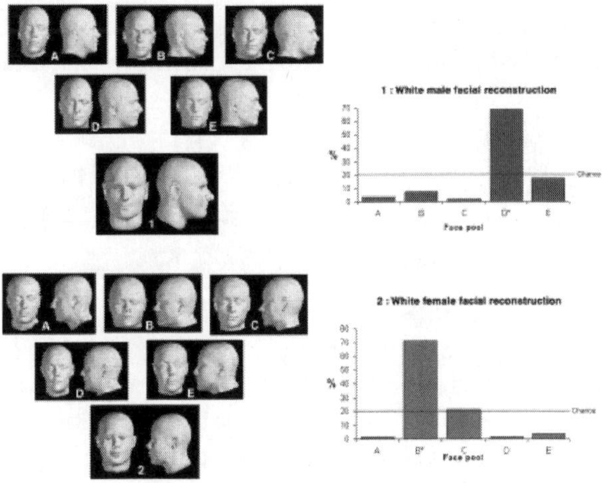

Fig. Face pool assessments of facial reconstruction accuracy. Top row: reconstruction of a white male skull (1) and a face pool (A–D). Bottom row: reconstruction of a white female skull (2) and a face pool (A–D). Graphs show the results of...

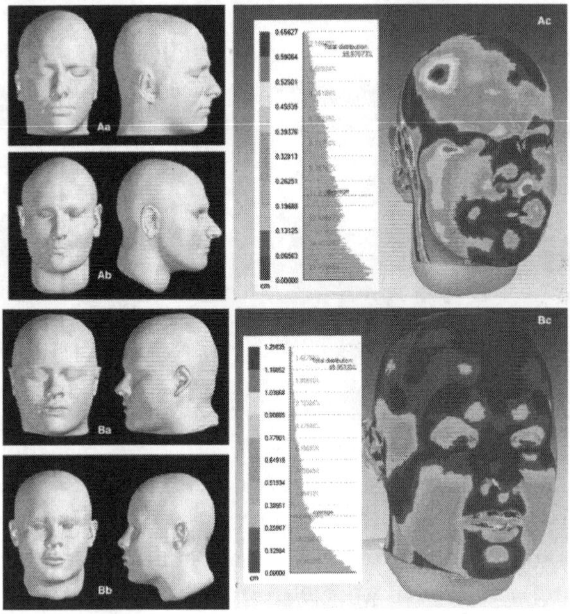

Fig. Morphometric assessment of facial reconstruction accuracy. The target white male (Aa), the facial reconstruction (Ab) and the contour map (Ac) showing the differences between the surfaces of the two faces in mm. The target white female (Ba), the facial reconstruction (Bb) and the contour map (Bc) showing the differences between the surfaces of the two faces in mm.

In conclusion, the laboratory studies of the Manchester method suggest that facial reconstruction can reproduce a sufficient likeness to allow recognition by a close friend or family member. It is not possible to produce a portrait and there are many details of the face that cannot be determined from the skull, but it should be possible to estimate the majority of facial feature morphology from skeletal detail.

ARTISTIC ANATOMY OR ANATOMICAL ART

The facial reconstruction technique involves three elements: the anatomical modelling, the morphology determination, and the depiction of the resulting face to the public.

Stage 1: Anatomical Modelling

Faces have, with some minor variation, a similar number of muscles with the same origins and attachments; indeed this is the very basis of anatomy education. However, the shape, size and relative position of these muscles will vary between individuals in relation to the bony matrix and some muscles of facial expression may be absent, duplicated or bifurcated. A routine visit to a dissecting room will show that there is considerable variation in anatomical structures and this human variation has long been recognized. Whilst some of these variations cannot easily be predicted (e.g. absence or bifurcation of a muscle of facial expression), skeletal assessment may indicate variation in muscle origins and/or attachments.

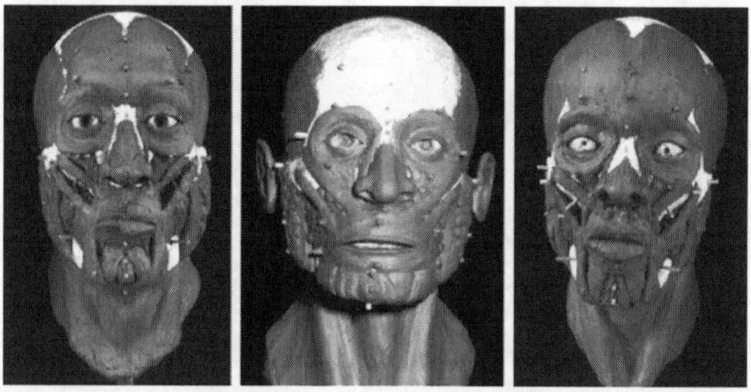

Fig. Three skulls with the addition of the muscle structure in modeling clay. Where the same muscles, with the same origins and insertions, are modeled onto three different skulls, clear differences can be observed in facial shape and proportions.

It is commonly reported in anatomical publications that some muscle attachments are associated with well defined areas on bone surfaces and research has established that muscle size and activity have a direct influence on craniofacial morphology.

The temporalis muscle attachment can often be seen on the surface of the temporal bone, the inferior border of the masseter muscle is frequently visible on the mandible and the zygomatic muscle origins may be visible at the zygomatic bones, that where the cheek bones are wide and heavy, the zygomaticus major and minor muscle origins will be positioned on the anterior surface of the zygomatic bone; in contrast, where the cheek bones are narrow and gracile, the origins of these two muscles will be positioned more laterally on the surface of the zygomatic bone. Awareness of these subtle differences means that sculpture of the facial musculature onto different skulls will produce different face shapes, with different proportions and different contours.

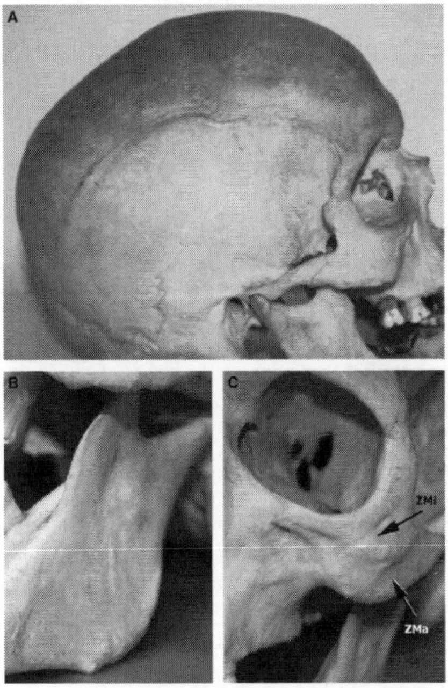

Fig. Muscle attachment markings for temporalis (A), masseter (B) and zygomatic (C) muscles. ZMi, zygomaticus minor; ZMa, zygomaticus major.

In conclusion, whilst sculptural skill is clearly useful, this stage of the reconstruction process should involve no artistic interpretation and, providing the muscles and other relevant structures (e.g. the parotid gland) are modelled following anatomical guidelines, the procedure will be both reproducible and reliable. Errors in muscle structure determination will occur at this stage of the reconstruction process, as the absence, duplication or bifurcation of facial expression muscles cannot be predicted from skeletal assessment, but the same errors should be produced by all practitioners. From teaching experience it is clear that this is the easiest stage of the reconstruction technique to learn, even with low levels of sculptural expertise.

Stage 2: Morphology Determination

The determination of facial features is carried out by assessing related bony detail. A number of standards are employed for each feature during this analysis, and many of the standards are related to anatomical principles. The determination of eye morphology is related to the position of the inner and outer canthi and the position of the eyeball in the orbit. Whitnall (1921) described the orbit in detail and many of our current standards rely on his extensive dissection study. The curves of the eyelid margins are not symmetrical and the upper lid is more pronounced than the lower, its height being greatest nearer the medial angle, whereas that of the lower lid is nearer the lateral angle (Whitnall, 1921).

The lateral canthal angle is more acute than the medial and lies in close contact with the globe, whereas the medial canthus extends towards the nose 5–7 mm away from the globe, being separated by the caruncula and the plica semilunaris. Merkel (1887) further described the radius of the upper curve as 16.5 mm and that of the lower as 22 mm. Whitnall (1912) described the position of the two canthi as being 'almost precisely determined, the inner by the naso-lacrymal duct (lacrimal fossa) and the outer by the slightly but definitely indicated malar tubercle'. Angel (1978) confirmed this and stated that the inner canthus can be placed 2 mm lateral to the lacrimal crest at its middle, and the outer canthus can be placed 3–4 mm medial to the malar tubercle.

Where the malar tubercle is absent, the outer canthus can be positioned 10 mm below the line of the zygomatico-frontal suture and 5–7 mm from the orbital margin. The centre point of the eyeball is determined as 2 mm below mid-orbit with the iris touching a tangent across the mid-supraorbital to mid-infraorbital bone (Whitnall, 1921). Whilst there has been disagreement in the past between practitioners regarding eyeball placement (Stephan & Davidson, 2008), current research results (Stephan, 2002; Wilkinson & Mautner, 2003) are in agreement with Whitnall and Wolff (summarized in Bron et al. 1997).

With these details determined and sculpture of the correct anatomical structures of the eye as described by Whitnall (1921) (a 24-mm diameter eyeball, a 12-mm diameter iris and the eyelids hugging the eyeball closely whilst clipping the edge of the iris as they cross the eyeball) there is little room for artistic interpretation.

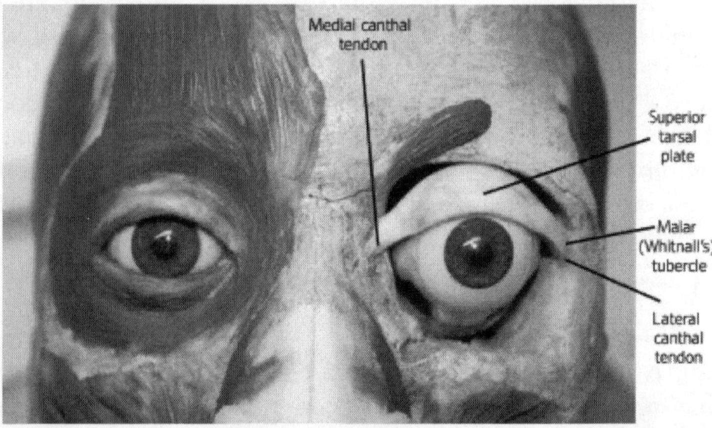

Fig. Anatomical reconstruction of the eye. Image courtesy of Caroline Needham, University of Dundee.

Traditionally the nose has been considered a feature with poor levels of reconstruction accuracy and there have been many studies assessing the relationship between the configuration of the nasal tissue with the bones surrounding the nasal aperture. It is anatomically predictable that the soft nose will be wider than the bony aperture, as a narrower soft nose would have no supporting structure. It also makes anatomical sense that the soft nose will not be very much wider than the bony aperture, as this would create a partial obstruction to the movement of air into the airways, forcing the air to change direction as it entered the nostrils, creating inefficient air passage.

Gerasimov (1955) suggested that the bony nasal aperture at its widest point will be three-fifths of the overall width of the soft nose, and this has been confirmed in a CT study of living subjects regardless of ethnic group (Rynn, 2006). Nasal base angle (angle between the upper lip and the columella) is determined by the direction of the nasal spine.

He stated that the axis of the nasal spine serves as a base for the soft nose and determination of nasal spine direction follows the point of the spine, as if it were an arrowhead. He also suggested that the end of the soft nose could be predicted as the point where a line following the projection of the last part of the nasal bones (at the rhinion) crosses a line following the direction of the nasal spine, and confirmed these standards with a blind study of 50 cadaver heads.

Gerasimov (1955) also claimed that the height of the upper border of the alae can be determined by the position of the crista conchalis and the profile of the nose is a mirror of the nasal aperture in profile. These standards have been confirmed using CT data of living subjects (Rynn, 2006); this study additionally confirmed previous papers suggesting that deviation of the nasal tip from the midline is associated with opposing nasal septum deviation (Selzter, 1944,Gray, 1965) and that nasal tip bifurcation is associated with a bifid nasal spine (Weaver & Bellinger, 1946).

Rynn's research also produced guidelines for nasal shape prediction, utilizing three cranial measurements that can be used to predict six soft nose measurements. When all these standards are applied to nasal morphology sculpture, there is little room for artistic interpretation, as illustrated by a blind study (Rynn, 2006; Rynn et al. 2008) using a sample of six skulls, where the predicted noses were compared with ante-mortem images of the faces, showing a high level of accuracy.

The morphology of the mouth is an area of the face where there is more reliance on artistic interpretation. Orthodontic and anatomic literature suggests that the form of the mouth is related to the occlusion of the teeth, the dental pattern (Subtelny, 1959) and the facial profile (Gerasimov, 1955). Where the upper teeth are more prominent than the lower teeth, the upper lip will be more prominent than the lower lip and vice versa, and different occlusion patterns will suggest different lip patterns (Gerasimov, 1955).

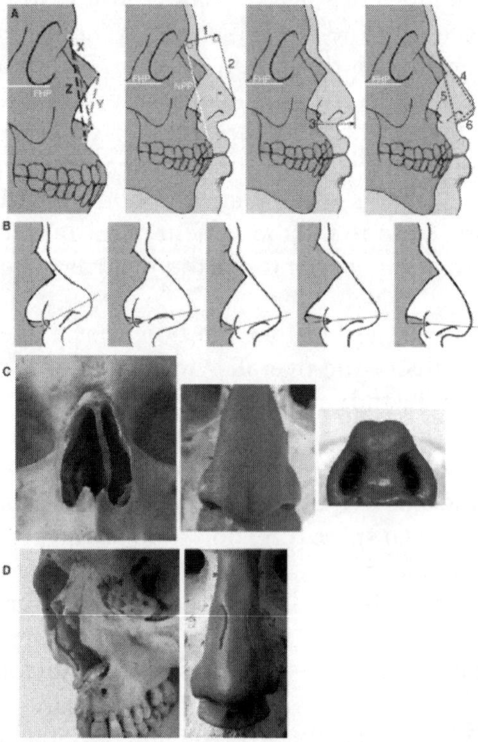

Fig. Anatomical sculpture of the nose (Rynn, 2006). Row A indicates the skeletal measurements (x,y,z) that provide soft tissue measurements 1–6. Row B demonstrates how the shape of the aperture in profile reflects the profile of the nose, and how the anterior nasal spine points to the nasal tip. Row C shows a bifurcated ANS and related bifid nose. Row D indicates how the undulating lateral borders of the nasal aperture are related to nasal shape.

There are some standards for determination of mouth shape, such as placement of the fissure at the mid-line of the maxillary incisor crowns (Angel, 1978) and the mouth corners on radiating lines from the first premolar-canine junction (Krogman & Iscan, 1986), or with intercanine distance as 75% of overall mouth width (Stephan & Henneberg, 2003), or the mouth corners positioned below the infraorbital foramina (Stephan & Murphy, 2008). There is also a positive correlation between upper lip thickness and maxillary enamel height, and between lower lip thickness and mandibular enamel height; sets of regression formulae can be utilized for White European and Indian subcontinent populations (Wilkinson et al. 2003).

However, the exact shape of the vermillion line is difficult to predict with any degree of accuracy, as illustrated by the results of a metrical evaluation in which lip shape was found to be one of the most error-prone areas of reconstruction (Wilkinson et al. 2006). Successful forensic reconstructions have been demonstrated where the practitioner has modeled the lips 'in sympathy' with the rest of the face (Prag & Neave, 1997), although this may be more luck than judgment. Ear shape is also very difficult to determine. Gerasimov (1955) considered the angle of ear to be parallel to the jaw line and stated that where the mastoid processes are directed downward, the earlobe will be attached (adherent), whereas where the mastoid processes point forward, the earlobe will be free. As yet very little information regarding ear shape, size and prominence can be determined reliably and typically standard ear casts will be attached to the reconstruction, which vary in relation to size and lobe pattern only (Wilkinson, 2004).

The final stage of the facial reconstruction process is the addition of a skin layer (with subdermal connective and adipose tissue) over the muscle structure to fill the face out to the level of the tissue depth pegs. The tissue depth pegs represent the mean tissue at an anatomical point related to the ethnic group, sex and age of the individual. There are a large number of international datasets including White European (Helmer, 1984), Indian (Sahni, 2002), North American Black, White and Hispanic (Manhein et al. 2000), South African Mixed Race, Japanese (Suzuki, 1948), Egyptian (El-Mehallawi & Soliman, 2001), Zulu (Auslebrook et al. 1996), Korean, Buryat, Kazakh, Bashhir, Uzbek, Armenian, Abkhazian, Russian and Lithuanian (Lebedinskaya et al. 1993) for adults, and White European (Wilkinson, 2002) and North American White, Black and Hispanic (Manhein et al. 2000) for juveniles. These datasets were measured on living individuals using ultrasound, MR or CT images.

There is an assumption with the use of these datasets that the individual has an average amount of fat over the surface of the face. This may or may not be true, but since it is currently impossible to determine facial fatness from the skeletal structure, this assumption is necessary and based on the hope that a familiar face will be recognizable even where the reconstruction involves a reduction or gain of weight. The skin layer follows the structure of the underlying muscles, so the main determinant of facial morphology is anatomical. Where the anatomical structures disagree with the tissue pegs, the tissue pegs will be removed, as the pegs only represent averages and will not be wholly appropriate for all skulls (Prag & Neave, 1997; Wilkinson, 2004). The addition of a skin layer does not in itself involve any degree of artistic interpretation, but the surface morphological detail is very important for both realistic depiction and artistic interpretation, and this will be related to the age of the individual.

Age-related changes to the face have been well documented and follow a predictable pattern (Gonzalez-Ulloa & Flores, 1965; Takema et al. 1994; Khalil et al. 1996): the skin loses elasticity due to biochemical changes in the underlying connective tissue that cause it to become less firmly attached to the underlying bone or muscles; wrinkles form due to changes in the distribution and formation of collagenous material in the skin, a decrease in the resilience of the fibres, and a decline in the number of fibroblasts leading to dehydration, sagging of flesh, loss of adipose tissue, blurring of iris detail, increased prominence of facial lines and hair loss. An old person may appear to have sunken eyes due to resorption of adipose tissue at the orbits and more visible veins beneath the thinner orbital skin, producing dark circles below the eyes (Gonzalez-Ulloa & Flores, 1965).

The suborbital region may also begin to sag, producing 'bags'. Nasolabial and mental creases will become more marked and deeper with increased age (Neave, 1998). Bone resorption at the alveolar processes with loss of teeth in later life will alter the jaw line and mouth significantly (Bodic et al. 2005). The nose and chin will appear more prominent and the distance between the nose and the chin will decrease, with the mouth appearing to sink into the face, and there is some growth of the cartilaginous portions of the nose and the ears throughout adulthood (Neave, 1998).

Although age-related changes to the skin surface follow a predictable pattern, the timing of this pattern is not predictable (Loth & Iscan, 1994; Novick, 1988; Orentreich, 1995). Age-related skin changes accrue more slowly in some people than others so there is a great deal of variation between individuals of the same age. Facial ageing is influenced by lifestyle and may be accelerated by external factors such as cigarette smoking, sleeping position, chronic alcohol consumption, sun damage, medication or loss of weight (Taister et al. 2000). These changes are also related to genetic factors, skin type, face shape and subcutaneous fat levels. Even if the accurate age of the individual is known, it is impossible to predict the surface texture of the face with any degree of reliability. This makes the reconstruction of the facial surface very difficult, with increased error related to increased age.

In conclusion, the majority of facial feature morphology (at least 67% according to the study byWilkinson et al. 2006) can be determined following scientific procedure, with lip and ear shapes the most difficult features to determine reliably. The skin layer in a child or young adult may be determined reliably, but as the adult age increases, the degree of artistic interpretation increases with respect to surface texture. From teaching experience it is clear that this is the most difficult stage of the reconstruction process to learn, and success is related to sculptural experience.

Stage 3: Depiction of the Face for Presentation to the Public

Forensic facial reconstructions may be presented to the general public using a variety of skin textures, colours, hairstyles and personal effects. Research suggests that different surface detail, such as hairstyle (Wright & Sladden, 2003), glasses and facial hair (Lewis & Johnson, 1997), can have an alarmingly strong effect upon recognition levels. It has also been shown that we find faces more difficult to recognize without surface detail and colour (Bruce et al. 1991). However, the amount of known surface detail regarding the identity of the individual will be different for each investigation. Although most forensic cases that involve facial reconstruction start with little information other than skeletal material, some scenes will reveal extra details regarding the appearance of the deceased, such as facial hair, skin colour, eye colour, hair or clothing.

Even where some details are known, the exact appearance of the individual feature will still be uncertain. For example, it may be known that the individual has white skin, but the range of white skin is wide, from milky-white and freckled to olive-toned. Similarly, the person may have dark brown, shoulder-length, straight hair, but the exact style or way that this was worn will be uncertain. These variations may have a great effect upon resemblance and recognition. In addition, characteristic facial appearance is often related to facial expression (Landau, 1989), which cannot be determined from the skeletal structure. Facial expression may also be inappropriate in forensic investigations or certain archaeological cases.

ARCHAEOLOGICAL INVESTIGATIONS

Archaeological investigations provide challenging facial analysis opportunities to depict faces from the past, to show the public how these ancient people looked and to enable comparison with contemporary faces (Prag & Neave, 1997). They may also provide different types of material compared to forensic investigations and the amount of artistic interpretation may be greater, depending on the level of preservation.

In forensic cases it is preferable to include only appearance details that have been ascertained directly from scene evidence, and not to estimate any unknowns, as incorrectly estimated details may confuse and discourage recognition and identification (Wilkinson, 2004). However, in archaeological investigations recognition of the face is rarely the primary objective and producing the most likely depiction may be more important than individual identity. In these cases the archaeologist will suggest the most probable hairstyle, hair colour, skin colour and eye colour from historical textual and pictorial evidence.

The depiction of preserved bodies

Ancient Egyptians have provided a rich source for analysis, as the mummification process preserves (with some modification) the soft tissues of the face as well as the skeletal material. The development of clinical imaging has allowed the non-invasive analysis of the soft and hard tissues of the faces of Ancient Egyptians. Cross-sectional data created by computed tomography (CT) can be employed to produce a three-dimensional digital model of the skull (Spoor et al. 2000) and replica skulls may be produced from digital data using stereolithography (Hjalgrim et al. 1995) or another form of three-dimensional model manufacture (Seitz et al. 2005). Digital 3D models of the skull can also be imported into computer-based facial reconstruction systems. Examples of such work include the facial reconstruction of Tut Ankh Amun (Gatliff, 2001; Handwerk, 2005), Nesperrenub (Taylor, 2004) and Janus (Tukker & Dassen, 1999). Often the assessment of mummified soft tissues will reveal details of facial morphology that cannot be determined from the skeleton alone, increasing the reliability and reducing the artistic interpretation of the facial reconstruction.

Frequently, the ears will be preserved and it may be possible to determine ear shape, size and detail directly from the mummy (Taylor, 2004). Similarly the hair line and lip pattern may be visible, details that require artistic interpretation in many facial reconstructions. Shrinkage, distortion and the effects of the mummification procedure must be considered, as these may affect the soft tissue appearance: many facial features will be distorted by the bandages, the nose may be distorted by brain removal and the eyes sunken due to post-mortem changes (Aufderheide, 2003). For example, the large majority of Egyptian mummies appear to exhibit hooked noses in profile, yet this is typically a bandage distortion rather than the actual profile of the nose, and experienced practitioners will recognize this phenomenon.

The facial appearance of bog bodies has been of great interest since the earliest discoveries, as the soft tissues are preserved by the acidity of the peat bog environment (Asingh & Lynnerup, 2007). This can lead to the incredible preservation of the faces of bog bodies, although there may be severe distortion due to the pressure of the burial environment. Where there is distortion of the soft tissues it can be difficult to visualize facial features and proportions, and in these circumstances facial reconstruction has been utilized (Prag & Neave, 1997, Chap. 8, Bergen et al. 2002). CT imaging allows visualization of the skull but the acidic conditions of the peat bog cause deterioration of the bone surface and this, in combination with any distortion, makes skeletal assessment for facial reconstruction difficult.

However, the lack of skeletal detail is more than compensated by the enormous amount of preserved facial morphology, and the facial reconstruction of bog bodies will be carried out primarily from the information provided by the soft tissue material (Wilkinson, 2007). As with mummies the ears will often be preserved and it may be possible to determine ear shape, size, protrusion and detail directly. Similarly, the vermillion line, hair line and facial wrinkle patterns may be visible, details that require artistic interpretation in many facial reconstructions (Wilkinson, 2009). Grauballe Man and Clonycavan Man are examples of such work (Wilkinson, 2007, 2009). Along with facial morphology detail, bog bodies often exhibit additional information regarding the hairstyle, hair colour and facial hair (Wilkinson, 2009), decreasing the artistic interpretation of the reconstruction further and increasing its reliability.

The depiction of disease and trauma

Archaeological cases can be challenging as the skeletal remains may exhibit certain pathological conditions, facial deformity or facial wounds. Where this is evident, the technique of facial reconstruction can be a valuable tool to help establish facial appearance relating to medical treatment, ancient disease processes and hereditary conditions. Where facial trauma was peri-mortem, it is usually inappropriate for the reconstruction to demonstrate any resulting wounds. Where the wounds show signs of healing it may be appropriate to demonstrate the soft tissue scarring as part of the reconstruction. The first facial reconstruction case that involved facial trauma was of Philip II Macedon from 250 BC ancient Greece (Prag & Neave, 1997), showing a well-healed wound to the right eye. A more recent example can be seen in the case of a soldier (number 16) from the Battle of Towton. His remains are part of the 1996 Towton Battle collection of 37 skeletons from the University of Bradford (Fiorato et al. 2000). The skull exhibited a well-healed blade injury to the left mandibular corpus. The blade wound had penetrated the lingual surface of the corpus, removing the apex of the first molar and impacting the internal surface of the right mandibular corpus.

A gaping hole remained between the left first and second molars where the bone had been removed. All margins of the wound were well healed and exhibited no evidence of infection. The reconstruction was produced demonstrating this healed wound, as it would have greatly altered the soldier's living facial appearance, with the resulting scar tissue distorting the left side of the face to a significant degree. In addition, he may have suffered eating and talking difficulties and some problems with mouth closure on the left side, leading to dribbling and a drooping of the left mouth corner (Wilkinson & Neave, 2003). Reconstructions depicting pathological conditions can provide useful information for comparison with contemporary populations and reconstructions have been produced of individuals with meningioma hyperostosis and a haemangioma (Wilkinson, 2008). These reconstructions provide more detail regarding the surface detail of the face, and in this way can be considered to involve increased scientific evaluation. However, the interpretation of trauma and disease, although based on science, may still be considered to include an element of artistic estimation.

The use of portraits for surface detail

Occasionally portraits may be available for use as reference material for the addition of surface detail, such as fatness, age-related changes, skin colour, eye colour, hairstyle and colour and facial hair. In these circumstances the resulting facial reconstructions can be considered more reliable in terms of resemblance. Usually the facial reconstruction process from skull to face is carried out blind to the portrait and then the surface detail is added to depict the face from the past more reliably.

Examples of the use of portraiture for surface detail are the facial reconstructions of Ancient Egyptian mummies with related portraits. The portraits (Fayuum portraits) all date from the 1st to 2nd centuries AD and were produced using an encaustic technique (Walker, 1997). These portraits have been analysed by Egyptologists to determine hairstyles, jewelry, fashion and social status. It has been estimated that there are more than 1000 mummy portraits, but fewer than 100 are still bound into their mummies. Several mummies with portraits have been studied (Wilkinson, 2003b; Prag, 2002; Brier & Wilkinson, 2005) and, in all cases, following a resemblance assessment, the portraits were used to apply additional surface detail to the reconstructions.

A recent study of the skull of Johann Sebastian Bach (1685–1750), the famous German composer and organist, included the assessment of a known portrait of Bach for additional surface information (Hansen, 2008). Bach was known to have sat for the painter Elias Gottlob Haussmann in 1746, a few years before his death (Müller, 1935). Haussmann (1695–1774) served as court painter at Dresden, and from 1720 as the official portraitist at Leipzig. The facial reconstruction was produced as a 3D computerized model using a laser scan of the bronze copy of the skull of Bach provided by the Bachhaus, Eisenach.

The skull suggested a middle-aged man with a large nose, strong chin and under-bite. Tissue depth data from contemporary German men in their 60s were employed for the reconstruction (Helmer, 1984), but the degree of fatness to the face was directly influenced by the Haussmann portrait. Records of Bach's life suggest that he had eye problems that left him with swollen, sore lids and the portrait depicted this affliction (David & Mendel, 1945; Zegers, 2005). The eye detail, eye colour, skin colour and facial ageing details were determined from the portrait (Fig. 10). The combination of skeletal assessment and portrait evaluation presents a facial depiction of Bach that can be considered as accurate as is possible with the material available.

TYPES OF RECONSTRUCTIONS

Two-Dimensional Reconstructions

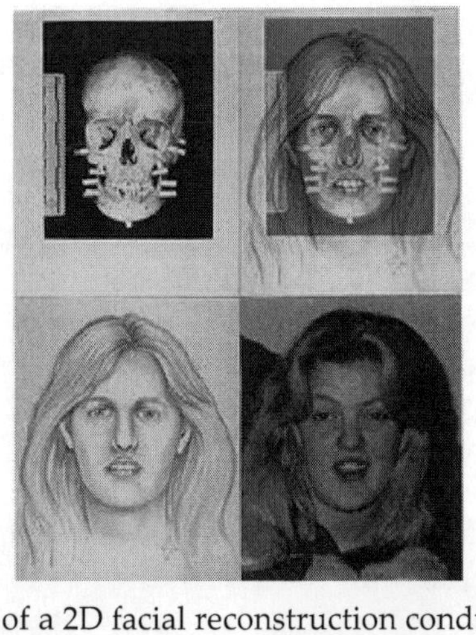

Fig. An example of a 2D facial reconstruction conducted by forensic artist Karen T. Taylor, and the subsequent identification.

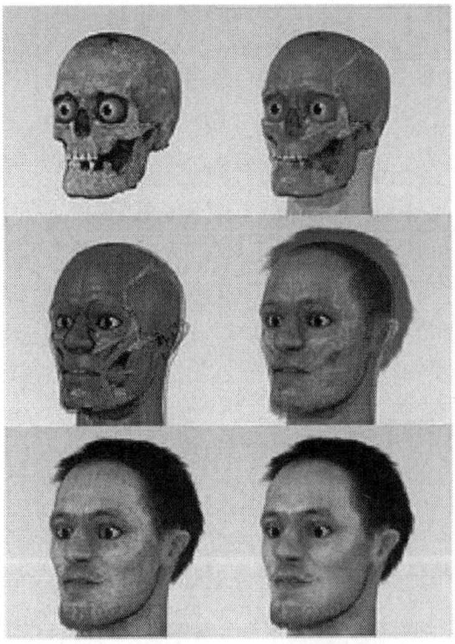

Fig. Forensic facial reconstruction of Alberto di Trento by Arc-Team and the 3D artist Cicero Moraes

Two-dimensional facial reconstructions are based on ante mortem photographs, and the skull. Occasionally skull radiographs are used but this is not ideal since many cranial structures are not visible or at the correct scale. This method usually requires the collaboration of an artist and a forensic anthropologist. A commonly used method of 2D facial reconstruction was pioneered byKaren T. Taylor of Austin, Texas during the 1980s. Taylor's method involves adhering tissue depth markers on an unidentified skull at various anthropological landmarks, then photographing the skull.

Life-size or one-to-one frontal and lateral photographic prints are then used as a foundation for facial drawings done on transparent vellum. Recently developed, the F.A.C.E. and C.A.R.E.S. computer software programs quickly produce two-dimensional facial approximations that can be edited and manipulated with relative ease. These programs may help speed the reconstruction process and allow subtle variations to be applied to the drawing, though they may produce more generic images than hand-drawn artwork.

Three-dimensional reconstructions

Three-dimensional facial reconstructions are either: 1) sculptures (made from casts of cranial remains) created with modeling clay and other materials or 2) high-resolution, three-dimensional computer images. Like two-dimensional reconstructions, three-dimensional reconstructions usually require both an artist and a forensic anthropologist. Computer programs create three-dimensional reconstructions by manipulating scanned photographs of the unidentified cranial remains, stock photographs of facial features, and other available reconstructions. These computer approximations are usually most effective in victim identification because they do not appear too artificial.

Superimposition

Superimposition is a technique that is sometimes included among the methods of forensic facial reconstruction. It is not always included as a technique because investigators must already have some kind of knowledge about the identity of the skeletal remains with which they are dealing (as opposed to 2D and 3D reconstructions, when the identity of the skeletal remains are generally completely unknown). Forensic superimpositions are created by superimposing a photograph of an individual suspected of belonging to the unidentified skeletal remains over an X-ray of the unidentified skull. If the skull and the photograph are of the same individual, then the anatomical features of the face should align accurately.

HISTORY

Welcker (1883) and His (1895) were the first to reproduce three-dimensional facial approximations from cranial remains. Most sources, however, acknowledge His as the forerunner in advancing the technique. His also produced the first data on average facial tissue thickness followed by Kollmann and Buchly who later collected additional data and compiled tables that are still referenced in most laboratories working on facial reproductions today. Facial reconstruction originated in two of the four major subfields of anthropology.

In biological anthropology, they were used to approximate the appearance of early hominid forms, while in archaeology they were used to validate the remains of historic figures. In 1964, Mikhail Gerasimov was probably the first to attempt paleo-anthropological facial reconstruction to estimate the appearance of ancient peoples

Although students of Gerasimov later used his techniques to aid in criminal investigations, it was Wilton M. Krogman who popularized facial reconstruction's application to the forensic field. Krogman presented his method for facial reconstruction in his 1962 book, detailing his method for approximation. Others who helped popularize three-dimensional facial reconstruction include Cherry (1977), Angel (1977), Gatliff (1984), Snow (1979), and Iscan (1986). In 2004 it was noted by Dr. Andrew Nelson of the University of Western Ontario, Department of Anthropology that noted Canadian artist Christian Corbet created the first forensic facial reconstruction of an approximate 2,200 year old mummy based on CT and laser scans. This reconstruction is known as the Sulman Mummy project.

TECHNIQUE FOR CREATING A THREE-DIMENSIONAL CLAY RECONSTRUCTION

Because a standard method for creating three-dimensional forensic facial reconstructions has not been widely agreed upon, multiple methods and techniques are used. The process detailed below reflects the method presented by Taylor and Angel from their chapter in Craniofacial Identification in Forensic Medicine, pgs 177-185. This method assumes that the sex, age, and race of the remains to undergo facial reconstruction have already been determined through traditional forensic anthropological techniques.

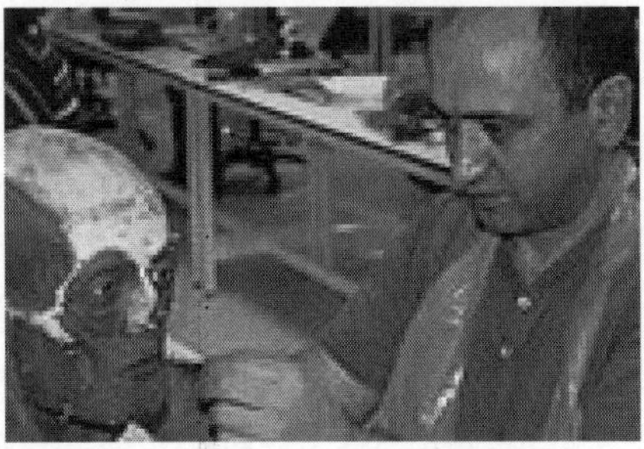

Fig. Forensic Facial Reconstruction at The University of Dundee-Scotland.

The skull is the basis of facial reconstruction; however, other physical remains that are sometimes available often prove to be valuable. Occasionally, remnants of soft tissue are found on a set of remains. Through close inspection, the forensic artist can easily approximate the thickness of the soft tissue over the remaining areas of the skull based on the presence of these tissues. This eliminates one of the most difficult aspects of reconstruction, the estimation of tissue thickness. Additionally, any other bodily or physical evidence found in association with remains (e.g. jewelry, hair, glasses, etc.) are vital to the final stages of reconstruction because they directly reflect the appearance of the individual in question.

Fig. Steps of forensic facial reconstruction of a mummy

Most commonly, however, only the bony skull and minimal or no other soft tissues are present on the remains presented to forensic artists. In this case, a thorough examination of the skull is completed. This examination focuses on, but is not limited to, the identification of any bony pathologies or unusual landmarks, ruggedness of muscleattachments, profile of the mandible, symmetry of the nasal bones, dentition, and wear of the occlusal surfaces. All of these features have an effect on the appearance of an individual's face.

Once the examination is complete, the skull is cleaned and any damaged or fragmented areas are repaired with wax. The mandible is then reattached, again with wax, according to the alignment of teeth, or, if no teeth are present, by averaging the vertical dimensions between the mandible and maxilla. Undercuts (like the nasal openings) are filled in with modeling clay and prosthetic eyes are inserted into the orbits centered between the superior and inferior orbitalrims. At this point, a plaster cast of the skull is prepared. Extensive detail of the preparation of such a cast is presented in the article from which these methods are presented.

After the cast is set, colored plastics or the colored ends of safety matches are attached at twenty-one specific "landmark" areas that correspond to the reference data. These sites represent the average facial tissue thickness for persons of the same sex, race, and age as that of the remains. From this point on, all features are added using modeling clay.

First, the facial muscles are layered onto the cast in the following order: temporalis, masseter, buccinator and occipito-frontals, and finally the soft tissues of the neck. Next, the nose and lips are reconstructed before any of the other muscles are formed. The lips are approximately as wide as the interpupillary distance.

However, this distance varies significantly with age, sex, race, and occlusion. The nose is one of the most difficult facial features to reconstruct because the underlying bone is limited and the possibility of variation is expansive. The nasal profile is constructed by first measuring the width of the nasal aperture and the nasal spine. Using a calculation of three times the length of the spine plus the depth of tissue marker number five will yield the approximate nose length.

Next, the pitch of the nose is determined by examining the direction of the nasal spine - down, flat, or up. A block of clay that is the proper length is then placed on the nasal spine and the remaining nasal tissue is filled in using tissue markers two and three as a guide for the bridge of the nose. The alae are created by first marking a point five millimeters below the bottom of the nasal aperture. After the main part of the nose is constructed the alae are created as small egg-shaped balls of clay, that are five millimeters in diameter at the widest point, these are positioned on the sides of the nose corresponding with the mark made previously. The alae are then blended to the nose and the overall structure of the nose is rounded out and shaped appropriately.

The muscles of facial expression and the soft tissue around the eyes are added next. Additional measurements are made according to race (especially for those with eye folds characteristic of Asian descent) during this stage. Next, tissues are built up to within one millimeter of the tissue thickness markers and the ears (noted as being extremely complicated to reproduce) are added. Finally, the face is "fleshed," meaning clay is added until the tissue thickness markers are covered, and any specific characterization is added (for example, hair, wrinkles in the skin, noted racial traits, glasses, etc.). The skull of Mozart was the basis of his facial reconstruction from anthropological data.

PROBLEMS WITH FACIAL RECONSTRUCTION

Insufficient tissue thickness data

There are multiple outstanding problems associated with forensic facial reconstruction. The most pressing issue relates to the data used to average facial tissue thickness. The data available to forensic artists are still very limited in ranges of ages, sexes, and body builds. This disparity greatly affects the accuracy of reconstructions. Until this data is expanded, the likelihood of producing the most accurate reconstruction possible is largely limited.

Lack of methodological standardization

A second problem is the lack of a methodological standardization in approximating facial features. A single, official method for reconstructing the face has yet to be recognized. This also presents major setback in facial approximation because facial features like the eyes and nose and individuating characteristics like hairstyle - the features most likely to be recalled by witnesses - lack a standard way of being reconstructed. Recent research on computer-assisted methods, which take advantage of digital image processing, pattern recognition, promises to overcome current limitations in facial reconstruction and linkage.

Subjectivity

Reconstructions only reveal the type of face a person may have exhibited because of artistic subjectivity. The position and general shape of the main facial features are mostly accurate because they are greatly determined by the skull.

FACIAL RECONSTRUCTION IN CULTURE

Due to the recent rise in popularity of television shows (e.g. CSI: Crime Scene Investigation, CSI: Miami, CSI: NY, NCIS, Bones, and the UK programme Meet the Ancestors) and feature films concerned with criminal investigations, forensics, and law enforcement, the presence of forensic facial reconstructions in the entertainment industry and the media has also increased. The way the fictional criminal investigators and forensic anthropologists utilize forensics and facial reconstructions are, however, often misrepresented (an influence known as the "CSI effect"). For example, the fictional forensic investigators will often call for the creation of a facial reconstruction as soon as a set of skeletal remains is discovered. In reality, facial reconstructions are widely used as a last resort to stimulate the possibility of identifying a victim.

Interest in discovering and identifying "lost" and/or significant historical figures has also increased the presence of forensic facial reconstruction among the public consciousness. For example, the facial reconstruction of Egypt's King Tut made the June 2005 cover of National Geographic Magazine.

Casdon Toys has produced a line of at-home facial reconstruction toys featuring Julius Caesar, King Tut, Queen Nefertiti, andNeanderthal Man. The popular CSI: Crime Scene Investigation television show has also produced an at-home facial reconstruction kit.

Recently facial reconstruction has been part of the process used by researchers attempting to identify human remains of two Canadian Army soldiers lost in World War I. One soldier was identified through DNA analysis in 2007, but due to DNA deterioration, identifying the second using the same techniques failed. In 2011, the second of the soldiers' remains discovered at Avion, Francewere identified through a combination of 3-D printing software, reconstructive sculpture and use of isotopic analysis of bone.

3

Human Identification Based on Cranial Computed Tomography Scan

An adult male, initially unidentified, passed away as a result of a traffic accident. The body was transported to the regional Legal Medicine Institute for identification and routine medico-legal examinations: determining the cause of death, its circumstances and the instrument that caused it. After 30 days, the victim remained unidentified and the body had not been claimed by any next of kin. As a result, routine proceedings were conducted prior to burial, such as taking facial photographs and collecting post-mortem fingerprints and DNA samples.

As the victim had been missing for a long period of time, the supposed family members searched for him in hospitals as well as with friends and relatives. The result of this search ultimately ended when the victim was recognized based on post-mortem photographs filed in the unclaimed bodies section of the Legal Medicine Institute where the body had been examined. Immediately, the standard fingerprints were requested, which should have been on file with the civil identification authorities. However, the file containing those fingerprints (taken when the victim was alive) could not be found, which made a comparison of the prints impossible. Therefore, another methodology for confirming the victim's identification became necessary, leading family members to search for information and documentation related to old medical and dental treatments as well as photographs of the missing person.

The results of this investigation revealed the existence of a brain CT scan and seven thoracic and abdominal conventional radiographs. The imaging exams had been made approximately 6 years prior to the victim's death as part of the diagnosis for skull trauma, resulting from a traffic accident at the time. As a result of trauma, part of the parietal bone and part of the squamous portion of the temporal bone, both on the left side, were surgically removed. On the right side of the cranium, in the parietal bone, the victim was subjected to a surgical intervention for intracranial decompression, at which time a circular-shaped osteotomy was performed. Afterwards, the victim did not undergo a new surgical procedure to replace the lost area; the defect remained uncovered by any sort of metal plate. During the evaluation of the CT scan images of the cranium, two well-defined hypodense images were observed from the scanogram (Figure 1a): the first with a polygonal shape, similar to a quadrilateral, situated in the left temporoparietal region (Figure 1b); and the second with a circular shape located in the right parietal region.

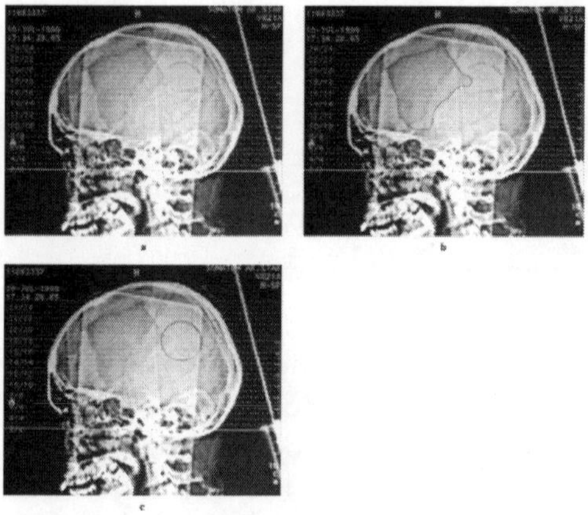

Fig. Scanogram demonstrating bilateral hypodense areas with different shapes (a) Delination of the edges of the images located on the left (b) and right (c) regions of the cranium

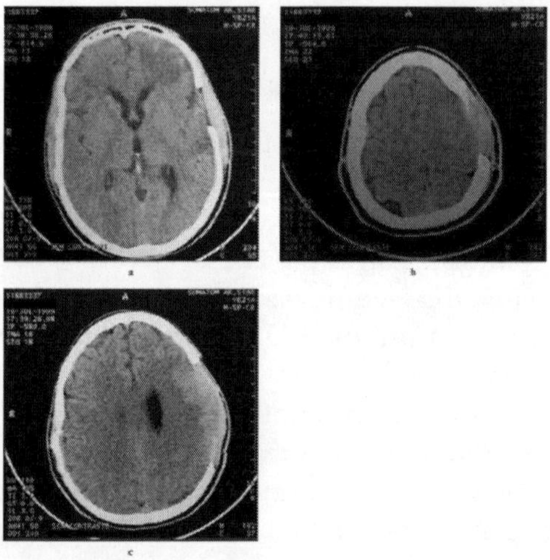

Fig. Axial slices of the cranium, evidencing bone window with the absence of part of the bone structure of the left side, extending from the squamous portion of the temporal (a) to the parietal bone (b). Also noticed is an irregularity in the surface of the right parietal bone at the same level of area of largest bone loss on the left side (c). This image also shows a subcortical haemorrhagic contusion

From the axial slice exam, an important bone loss was observed in the left side of the cranium, extending from the beginning of the squamous portion of temporal bone to the parietal bone. It was also possible to observe the borders of a circular-shaped fenestration in the right parietal bone at the level of the largest axis of bone loss on the left side.

Given this new possibility for identification, the body was exhumed approximately 1 year after burial, at which point only the cranium and the mandible were collected. The parts were adequately cleaned and bone fragments and teeth were glued, which enabled the performance of the post-mortem exams.

By examining the cranium, a cut in the cranial vault was detected, compatible with that produced during the autopsy. A circular-shaped fenestration in the right parietal region and a polygonal-shaped open defect in the left temporoparietal region were also detected (Figure 3) — both with dimensions, shapes and locations compatible with the images from the CT scan made in vivo. The bone edges that delimited these two bone lesions appeared blunt and smooth, indicating they had been produced in life, as a result of trauma (left side) and surgical procedure (right side), with no signs of recent change and not related to the medico-legal procedures performed.

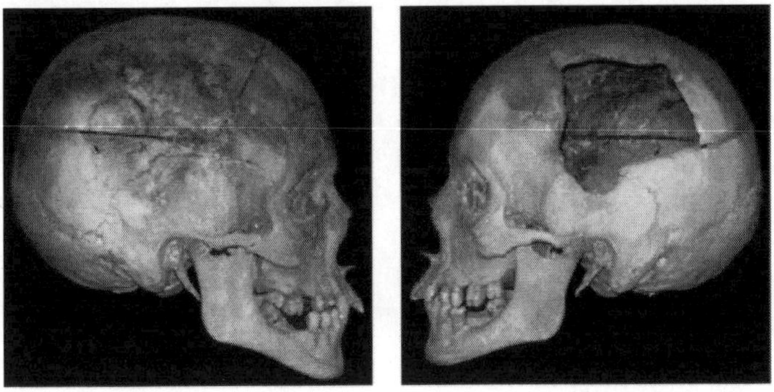

a b

Fig. Left (a) and right (b) lateral view of the cranium, reconstituted after exumation

DISCUSSION

The identification of human remains from the direct comparison of morpho-radiographic traits evidenced through in vivo and post-mortem images has become routine in forensic medicine departments. Of these, the radiographs most often used for forensic comparison are dental,1 cranial2 and extremities. Nevertheless, with the increased use of CT scans for clinical uses, these examinations have also become a relevant source of information for human identification, as they are currently performed at most public or private medical facilities.

Regarding the specific case presented here, the positive identification of the victim was possible because a cranial CT scan that had been made when the victim was alive was located, which allowed the visualization and measurements of specific characteristics related to bone loss owing to skull trauma resulting in surgical intervention. These traits were located on the left and right sides of the cranium. The qualitative and quantitative aspects of these findings, regarding location, borders and dimensions, were also adequately evidenced in the post-mortem examination, producing extremely relevant results that allowed a positive identification of the victim to be established exclusively by CT scan analysis.

The result of the analysis of thoracic and abdominal conventional radiographs, when compared with the post-mortem findings, did not show specific features that could positively identify the victim, demonstrating that the images of the craniofacial region had a higher potential for identification when compared with the thoracico-abdominal region.

Highlights must also be given to the fact that certain legal medicine facilities already have CT scan equipment that is used exclusively for cadaver examinations. This demonstrates the importance of investing in imaging technology for forensic needs. In this regard, the reference is the Legal Medicine Institute of Copenhagen, Denmark, which since 2002 routinely performs a full CT scan of all corpses examined in the forensic pathology department prior to the traditional autopsy.

The use of CT scans before classical autopsy are relevant to the analysis of trauma, decomposed bodies12 or human remains, helping or even making possible the identification of the circumstances and cause of death, and also in enhancing the view of anatomical features that could be disguised. A post-mortem CT scan also significantly contributes to image comparison in human identification because it represents an exam of quality, objectivity, accuracy and data which has general acceptance by the scientific community.

It is of upmost importance that imaging examinations produced during medical appointments at private practices, clinics or hospitals (public or private) be filed correctly. In case these exams are not filed, they should preferably be given to the patients or their legal guardians, emphasizing the importance of careful storage in case they become necessary. To summarize, CT scans produced in life, in addition to their clinical importance, can currently be used as an aiding tool for the identification of unidentified bodies, producing reliable results and important evidence to the authorities.

COMPUTED TOMOGRAPHY OF THE HEAD

Computed tomography (CT) of the head or Computed Axial Tomography (CAT) scanning uses a series of x-rays of the head taken from many different directions. Typically used for quickly viewing brain injuries, CT scanning uses a computer program that performs a numerical integral calculation (the inverse

Radon transform) on the measured x-ray series to estimate how much of an x-ray beam is absorbed in a small volume of the brain. Typically the information is presented as a series of cross sections of the brain. In approximation, the denser a material is, the whiter a volume of it will appear on the scan, just as in the more familiar "flat" X-rays, where dense bone appears white. CT scans are primarily used for evaluating swelling from tissue damage in the brain, bleeding, and in assessment of ventricle size.

Modern CT scanning can provide reasonably good images in a matter of minutes. A spiral CT scan of the head may be performed in 10–30 seconds, making it a good option for children and adults with difficulty holding still for longer periods. The use of contrast dye in CT angiography provides good visualization of the vascular structures and leaks in the blood vessels.

USES

Computed tomography (CT) has become the diagnostic modality of choice for head trauma due to its accuracy, reliability, safety, and wide availability. The changes in microcirculation, impaired auto-regulation, cerebral edema, and axonal injury start as soon as head injury occurs and manifest as clinical, biochemical, and radiological changes. Proper therapeutic management of brain injury is based on correct diagnosis and appreciation of the temporal course of the disease process. CT scan detects and precisely localizes the intracranial hematomas, brain contusions, edema and foreign bodies. Because of the widespread availability of CT, there is reduction in arteriography, surgical intervention and skull radiography.

ORBITAL VIEWS FOR EYE-RELATED DISORDERS

Special views focusing on the orbit of the eye may be taken to investigate concerns relating to the eye. CT scans are used by physicians specializing in treating the eye (ophthalmologists) to detect foreign bodies (especially metallic objects), fractures, abscesses, cellulitis, sinusitis, bleeding within the skull (intracranial bleeding), proptosis,Graves disease changes in the eye, and evaluation of the orbital apex and cavernous sinus.

COMPARISON WITH MRI

MRI (magnetic resonance imaging) provides more sensitivity in the evaluation of the cavernous sinus and the orbital apex. One advantage over a brain MRI is in the evaluation of intracerebral calcifications.

Cautions

Several different views of the head are available, including axial, coronal, reformatted coronal, and reformatted sagittal images. However, coronal images require the patienthyperextend their neck, which must be avoided if any possibility of neck injury exists.

X-RAY COMPUTED TOMOGRAPHY

X-ray computed tomography (x-ray CT) is a medical imaging procedure that uses computer-processed x-rays to producetomographic images or 'slices' of specific areas of the body. These cross-sectional images are used for diagnostic and therapeutic purposes in various medical disciplines. Digital geometry processing is used to generate a three-dimensionalimage of the inside of an object from a large series of two-dimensional X-ray images taken around a single axis of rotation.

Because x-ray CT is the most common form of CT in medicine and various other contexts, the term computed tomographyalone (or CT) is often used to refer to x-ray CT, although other types exist (such as single-photon emission computed tomography). Older, less preferred terms that also refer to x-ray CT are computed axial tomography (CAT scan) andcomputer-assisted tomography. X-ray CT is a form of radiography, although the word "radiography" used alone usually refers, by wide convention, to nontomographic radiography.

CT produces a volume of data that can be manipulated in order to demonstrate various bodily structures based on their ability to block the X-ray beam. Although, historically, the images generated were in the axial or transverse plane, perpendicular to the long axis of the body, modern scanners allow this volume of data to be reformatted in various planes or even as volumetric (3D) representations of structures.

Although most common in medicine, CT is also used in other fields, such as nondestructive materials testing. Another example is archaeological uses such as imaging the contents of sarcophagi. Individuals responsible for performing CT exams are called Radiologic Technologists or Radiographers and are required to be licensed in most states.

Usage of CT has increased dramatically over the last two decades in many countries. An estimated 72 million scans were performed in the United States in 2007. One study estimated that as many as 0.4% of current cancers in the United States are due to CTs performed in the past and that this may increase to as high as 1.5 to 2% with 2007 rates of CT usage; however, this estimate is disputed. Kidney problems following intravenous contrast agents may also be a concern in some types of studies.

DIAGNOSTIC USE

Since its introduction in the 1970s, CT has become an important tool in medical imaging to supplement X-rays and medical ultrasonography. It has more recently been used for preventive medicine or screening for disease, for example CT colonography for patients with a high risk of colon cancer, or full-motion heart scans for patients with high risk of heart disease. A number of institutions offer full-body scans for the general population although this practice goes against the advice and official position of many professional organizations in the field.

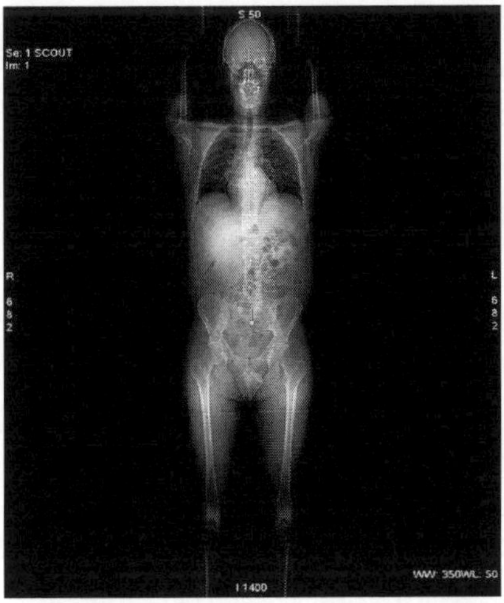

Fig. Picture of a CT scout (aka scanogramor topogram) as used for planning every scan slice.

Head

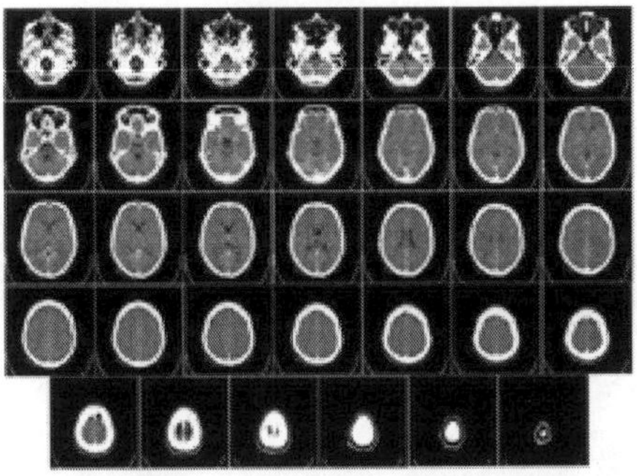

Fig. Computed tomography of human brain, from base of the skull to top. Taken with intravenous contrast medium.

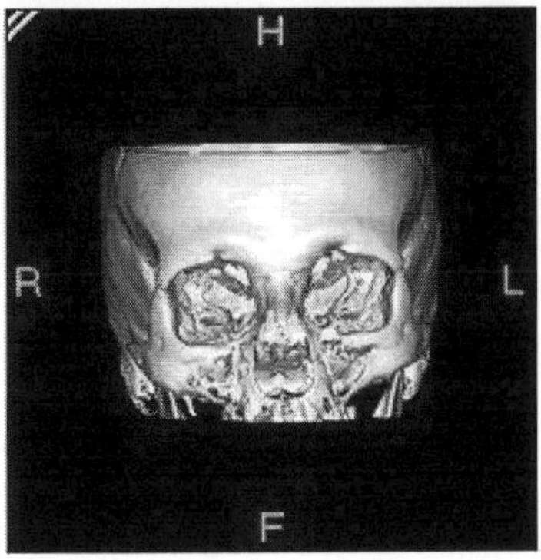

Fig. Bone reconstructed in 3D

CT scanning of the head is typically used to detect infarction, tumors, calcifications, haemorrhage and bone trauma. Of the above, hypodense (dark) structures can indicate edema and infarction, hyperdense (bright) structures indicate calcifications and haemorrhage and bone trauma can be seen as disjunction in bone windows. Tumors can be detected by the swelling and anatomical distortion they cause, or by surrounding edema. Ambulances equipped with small bore multi-sliced CT scanners respond to cases involving stroke or head trauma.

Lungs

CT can be used for detecting both acute and chronic changes in the lung parenchyma, that is, the internals of the lungs. It is particularly relevant here because normal two-dimensional X-rays do not show such defects. A variety of techniques are used, depending on the suspected abnormality. For evaluation of chronic interstitial processes (emphysema, fibrosis, and so forth), thin sections with high spatial frequency reconstructions are used; often scans are performed both in inspiration and expiration. This special technique is called high resolution CT. Therefore, it produces a sampling of the lung and not continuous images.

Pulmonary angiogram

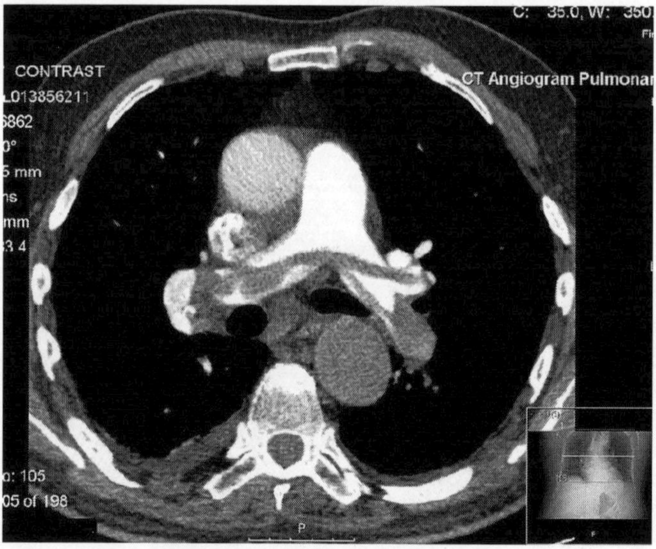

Fig. Example of a CTPA, demonstrating a saddle embolus (dark horizontal line) occluding the pulmonary arteries (bright white triangle)

CT pulmonary angiogram (CTPA) is a medical diagnostic test used to diagnose pulmonary embolism (PE). It employs computed tomography and an iodine based contrast agent to obtain an image of the pulmonary arteries.

Cardiac

With the advent of subsecond rotation combined with multi-slice CT (up to 320-slices), high resolution and high speed can be obtained at the same time, allowing excellent imaging of the coronary arteries (cardiac CT angiography).

Abdominal and pelvic

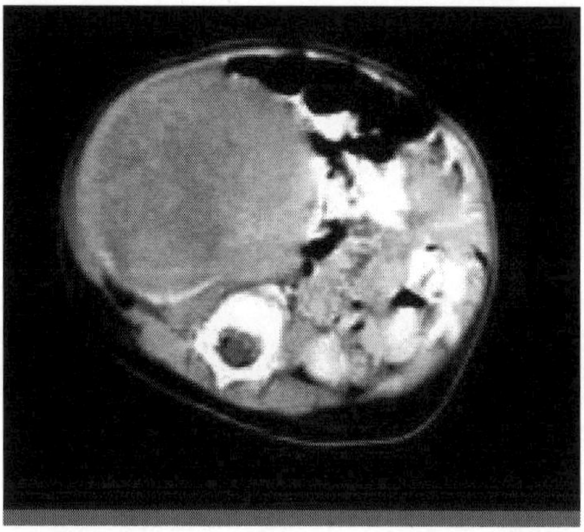

Fig. CT Scan of 11 cm Wilms' tumor of right kidney in 13 month old patient.

CT is a sensitive method for diagnosis of abdominal diseases. It is used frequently to determine stage of cancer and to follow progress. It is also a useful test to investigate acute abdominal pain.

Extremities

CT is often used to image complex fractures, especially ones around joints, because of its ability to reconstruct the area of interest in multiple planes. Fractures, ligamentous injuries and dislocations can easily be recognised with a 0.2 mm resolution.

ADVANTAGES

There are several advantages that CT has over traditional 2D medical radiography. First, CT completely eliminates the superimposition of images of structures outside the area of interest. Second, because of the inherent high-contrast resolution of CT, differences between tissues that differ in physical density by less than 1% can be distinguished. Finally, data from a single CT imaging procedure consisting of either multiple contiguous or one helical scan can be viewed as images in the axial, coronal, or sagittal planes, depending on the diagnostic task. This is referred to as multiplanar reformatted imaging.

CT is regarded as a moderate- to high-radiation diagnostic technique. The improved resolution of CT has permitted the development of new investigations, which may have advantages; compared to conventional radiography, for example, CT angiography avoids the invasive insertion of a catheter. CT Colonography (also known as Virtual Colonoscopy or VC for short) may be as useful as a barium enema for detection of tumors, but may use a lower radiation dose. CT VC is increasingly being used in the UK as a diagnostic test for bowel cancer and can negate the need for a colonoscopy.

The radiation dose for a particular study depends on multiple factors: volume scanned, patient build, number and type of scan sequences, and desired resolution and image quality. In addition, two helical CT scanning parameters that can be adjusted easily and that have a profound effect on radiation dose are tube current and pitch. Computed tomography (CT) scan has been shown to be more accurate than radiographs in evaluating anterior interbody fusion but may still over-read the extent of fusion.

ADVERSE EFFECTS

Cancer

The ionizing radiation in the form of x-rays used in CT scans are energetic enough to directly or indirectly damage DNA. This and other types of DNA damage are occasionally not corrected properly by cellular repair mechanisms. Such damage to the DNA occasionally leads to cancer. The estimates of harm from CT are partly based on similar radiation exposures experienced by those present during the atomic bomb explosions in Japan during the second world war and those of nuclear industry workers.

There is a small increased risk of cancer with CT scans. It is estimated that 0.4% of current cancers in the United States are due to CTs performed in the past and that this may increase to as high as 1.5–2% with 2007 rates of CT usage; however, this estimate is disputed. This would be equivalent to one in 1000 to one in 2000 increased risk of developing a fatal cancer per 10 mSv CT scan, or 29,000 new cancer cases in the United States due to the number of scans done in 2007 and 2100 new cancers in the United Kingdom. This additional risk is still low compared to the background risk of dying from cancer of ~20%. The most common cancers caused by radiation exposure are thought to be lung cancer, breast cancer, thyroid cancer, stomach cancer andleukemia.

A person's age plays a significant role in the subsequent risk of cancer. Estimated lifetime cancer mortality risks from an abdominal CT of a 1-year-old is 0.1% or 1:1000 scans. The risk for someone who is 40 years old is half that of someone who is 20 years old with substantially less risk in the elderly. The International Commission on Radiological Protection estimates that the risk to a fetus being exposed to 10 mGy (a unit of radiation exposure, see Gray (unit)) increases the rate of cancer before 20 years of age from 0.03% to 0.04% (for reference a CT pulmonary angiogram exposes a fetus to 4 mGy). A 2012 review did not find an association between medical radiation and cancer risk in children noting however the existence of limitations in the evidences over which the review is based.

CT scans can be performed with different settings for lower exposure in children with most manufacturers of CT scans as of 2007 having this function built in. Furthermore, certain conditions can require children to be exposed to multiple CT scans. Studies support informing parents of the risks of pediatric CT scanning.

Contrast

In the United States half of CT scans involve intravenously injected radiocontrast agents. The most common reactions from these agents are mild, including nausea, vomiting and an itching rash; however, more severe reactions may occur. Overall reactions occur in 1 to 3% with nonionic contrast and 4 to 12% of people with ionic contrast.

Skin rashes may appear within a week to 3% of people. The old radiocontrast agents caused anaphylaxis in 1% of cases while the newer, lower-osmolar agents cause reactions in 0.01–0.04% of cases. Death occurs in about two to 30 people per 1,000,000 administrations with newer agents being safer. When deaths do occur it is more typically in those who are female, elderly or in poor health and is secondary to either anaphylaxis or acute renal failure.

The contrast agent may induce contrast-induced nephropathy. This occurs in 2 to 7% of people who receive these agents, with greater risk in those who have preexistingrenal insufficiency, preexisting diabetes, or reduced intravascular volume.

People with mild kidney impairment are usually advised to ensure full hydration for several hours before and after the injection. For moderate kidney failure, the use of iodinated contrast should be avoided; this may mean using an alternative technique instead of CT.

Those with severe renal failure requiring dialysis require less strict precautions, as their kidneys have so little function remaining that any further damage would not be noticeable and the dialysis will remove the contrast agent; it is normally recommended, however, to arrange dialysis as soon as possible following contrast administration to minimize any adverse effects of the contrast.

In addition to the use of intravenous contrast, orally administered contrast agents are frequently used when examining the abdomen. These are frequently the same as the intravenous contrast agents, merely diluted to approximately 10% of the concentration.

However, oral alternatives to iodinated contrast exist, such as very dilute (0.5–1% w/v)barium sulfate suspensions. Dilute barium sulfate has the advantage that it does not cause allergic-type reactions or kidney failure, but cannot be used in patients with suspected bowel perforation or suspected bowel injury, as leakage of barium sulfate from damaged bowel can cause fatal peritonitis.

SCAN DOSE

Examination	Typical Effective dose (mSv) to the Whole Body	Typical Absorbed dose (mGy) to the Organ in Question
Annual background radiation	2.4	2.4
Chest X-ray	0.02	0.01–0.15
Head CT	1–2	56
Screening mammography	0.4	3
Abdomen CT	8	14
Chest CT	5–7	13
CT colonography	6–11	
Chest, abdomen and pelvis CT	9.9	12
Cardiac CT angiogram	9–12	40–100
Barium enema	15	15
Neonatal abdominal CT	20	20

The table reports average radiation exposures, however, there is a wide variation in radiation doses between similar scan types, where the highest dose could be as much as 22 times higher than the lowest dose. A typical plain film x-ray involves radiation dose of 0.01 to 0.15 mGy, while a typical CT can involve 10–20 mGy for specific organs, and can go up to 80 mGy for certain specialized CT scans.

For purposes of comparison, the world average dose rate from naturally occurring sources of background radiation is 2.4 mSv per year, equal for practical purposes in this application to 2.4 mGy per year. While there is some variation, most people (99%) received less than 7 mSv per year as background radiation. Medical imaging as of 2007 accounted for half of the radiation exposure of those in the United States with CT scans making up two thirds of this amount. In the United Kingdom it accounts for 15% of radiation exposure. The average radiation dose from medical sources is ~0.6 mSv per person globally as of 2007. Those in the nuclear industry in the United States are limited to doses of 50 mSv a year and 100 mSv every 5 years.

Radiation dose units

The radiation dose reported in the gray or mGy unit is proportional to the amount of energy that the irradiated body part is expected to absorb, and the physical effect (such as DNA double strand breaks) on the cells' chemical bonds by x-ray radiation is proportional to that energy.

The sievert unit is used in the report of the effective dose. The sievert unit in the context of CT scans, does not correspond to the actual radiation dose that the scanned body part absorbs, but rather to another radiation dose of another scenario, in which the whole body absorbs the other radiation dose, and where the other radiation dose is of a magnitude that is estimated to have the same probability to induce cancer as the CT scan. Thus, as is shown in the table above, the actual radiation that is absorbed by a scanned body part is often much larger than the effective dose suggests. A specific measure, termed the computed tomography dose index (CTDI), is commonly used as an estimate of the radiation absorbed dose for tissue within the scan region, and is automatically computed by medical CT scanners.

The equivalent dose is the effective dose of a case, in which the whole body would actually absorb the same radiation dose, and the sievert unit is used in its report. In the case of non-uniform radiation, or radiation given to only part of the body, which is common for CT examinations, using the local equivalent dose alone would overstate the biological risks to the entire organism.

Excess doses

In October, 2009, the US Food and Drug Administration (FDA) initiated an investigation of brain perfusion CT (PCT) scans, based on overdoses of radiation caused by incorrect settings at one particular facility for this particular type of CT scan. Over 256 patients over an 18 month period were exposed, over 40% lost patches of hair, and prompted the editorial to call for increased CT quality assurance programs, while also noting that "while unnecessary radiation exposure should be avoided, a medically needed CT scan obtained with appropriate acquisition parameter has benefits that outweigh the radiation risks." Similar problems have been reported at other centers. These incidents are believed to be due to human error.

Campaigns

In response to increased concern by the public and the ongoing progress of best practices, The Alliance for Radiation Safety in Pediatric Imaging was formed within theSociety for Pediatric Radiology. In concert with The American Society of Radiologic Technologists, The American College of Radiology and The American Association of Physicists in Medicine, the Society for Pediatric Radiology developed and launched the Image Gently Campaign which is designed to maintain high quality imaging studies while using the lowest doses and best radiation safety practices available on pediatric patients. This initiative has been endorsed and applied by a growing list of various professional medical organizations around the world and has received support and assistance from companies that manufacture equipment used in Radiology.

Following upon the success of the Image Gently campaign, the American College of Radiology, the Radiological Society of North America, the American Association of Physicists in Medicine and the American Society of Radiologic Technologists have launched a similar campaign to address this issue in the adult population called Image Wisely. The World Health Organization and

International Atomic Energy Agency (IAEA) of the United Nations have also been working in this area and have ongoing projects designed to broaden best practices and lower patient radiation dose.

PREVALENCE

Usage of CT has increased dramatically over the last two decades. An estimated 72 million scans were performed in the United States in 2007. Of these, six to eleven percent are done in children, an increase of seven to eightfold from 1980. Similar increases have been seen in Europe and Asia. In Calgary, Canada 12.1% of people who present to the emergency with an urgent complaint received a CT scan, most commonly either of the head or of the abdomen. The percentage who received CT, however, varied markedly by the emergency physician who saw them from 1.8% to 25%. In the emergency department in the United States, CT or MRI imaging is done in 15% of people who present with injuries as of 2007 (up from 6% in 1998).

The increased use of CT scans has been the greatest in two fields: screening of adults (screening CT of the lung in smokers, virtual colonoscopy, CT cardiac screening, and whole-body CT in asymptomatic patients) and CT imaging of children. Shortening of the scanning time to around 1 second, eliminating the strict need for the subject to remain still or be sedated, is one of the main reasons for the large increase in the pediatric population (especially for the diagnosis of appendicitis). As of 2007 in the United States a proportion of CT scans are performed unnecessarily. Some estimates place this number at 30%. There are a number of reasons for this including: legal concerns, financial incentives, and desire by the public. There is also otherwise healthy people receiving full body CT scans as screening for which there is no evidence to support.

PROCESS

X-ray slice data is generated using an X-ray source that rotates around the object; X-ray sensors are positioned on the opposite side of the circle from the X-ray source. The earliest sensors were scintillation detectors, with photomultiplier tubes excited by (typically) cesium iodide crystals. Cesium iodide was replaced during the 1980s by ion chambers containing high-pressure Xenongas. These systems were in turn replaced by scintillation systems based on photodiodes instead of photomultipliers and modern scintillation materials (for example rare earth garnet or rare earth oxide ceramics) with more desirable characteristics.

Initial machines would rotate the X-ray source and detectors around a stationary object. Following a complete rotation, the object would be moved along its axis, and the next rotation started. Newer machines permitted continuous rotation with the object to be imaged slowly and smoothly slid through the X-ray ring. These are called helical or spiral CT machines. A subsequent development of helical CT was multi-slice (or multi-detector) CT; instead of a single row of detectors, multiple rows of detectors are used effectively capturing multiple cross-sections simultaneously. Systems with a very large number of detector rows, such that the z-axis coverage is comparable to the xy-axis coverage are often termed cone beam CT, due to the shape of the X-ray beam (strictly, the beam is pyramidal in shape, rather than conical).

In conventional CT machines, an X-ray tube and detector are physically rotated behind a circular shroud (see the image above right). An alternative, short lived design, known as electron beam tomography (EBT), used electromagnetic deflection of an electron beam within a very large conical X-ray tube and a stationary array of detectors to achieve very high temporal resolution, for imaging of rapidly moving structures, for example the coronary arteries. Cone-beam CT functionality is also an increasingly common function found in medical fluoroscopy equipment; by rotating the fluoroscope around the patient, a geometry similar to CT can be obtained, and by treating the 2D X-ray detector in a manner similar to a CT detector with a massive number of rows, it is possible to reconstruct a 3D volume from a single rotation using suitable software.

CT is used in medicine as a diagnostic tool and as a guide for interventional procedures. Sometimes contrast materials such asintravenous iodinated contrast are used. This is useful to highlight structures such as blood vessels that otherwise would be difficult to delineate from their surroundings. Using contrast material can also help to obtain functional information about tissues.

A visual representation of the raw data obtained is called a sinogram, yet it is not sufficient for interpretation. Once the scan data has been acquired, the data must be processed using a form of tomographic reconstruction, which produces a series of cross-sectional images. In terms of mathematics, the raw data acquired by the scanner consists of multiple "projections" of the object being scanned. These projections are effectively the Radon transformation of the structure of the object. Reconstruction, essentially involves solving the inverse Radon transformation.

The technique of filtered back projection is one of the most established algorithmic techniques for this problem. It is conceptually simple, tunable and deterministic. It is also computationally undemanding, with modern scanners requiring only a few milliseconds per image. However, this is not the only technique available: the original EMI scanner solved the tomographic reconstruction problem by linear algebra, but this approach was limited by its high computational complexity, especially given the computer technology available at the time. More recently, manufacturers have developed iterative physical model-based maximum likelihoodexpectation maximization techniques. These techniques are advantageous because they use an internal model of the scanner's physical properties and of the physical laws of X-ray interactions. Earlier methods, such as filtered back projection, assume a perfect scanner and highly simplified physics, which leads to a number of artifacts, high noise and impaired image resolution. Iterative techniques provide images with improved resolution, reduced noise and fewer artifacts, as well as the ability to greatly reduce the radiation dose in certain circumstances. The disadvantage is a very high computational requirement, but advances in computer technology and high-performance computing techniques, such as use of highly parallel GPU algorithms, now allow practical use.

Pixels in an image obtained by CT scanning are displayed in terms of relative radiodensity. The pixel itself is displayed according to the mean attenuation of the tissue(s) that it corresponds to on a scale from +3071 (most attenuating) to -1024 (least attenuating) on the Hounsfield scale. Pixel is a two dimensional unit based on the matrix size and the field of view. When the CT slice thickness is also factored in, the unit is known as aVoxel, which is a three-dimensional unit. The phenomenon that one part of the detector cannot differentiate between different tissues is called the "Partial Volume Effect".

That means that a big amount of cartilage and a thin layer of compact bone can cause the same attenuation in a voxel as hyperdense cartilage alone. Water has an attenuation of 0 Hounsfield units (HU), while air is -1000 HU, cancellous bone is typically +400 HU, cranial bone can reach 2000 HU or more (os temporale) and can cause artifacts. The attenuation of metallic implants depends on atomic number of the element used: Titanium usually has an amount of +1000 HU, iron steel can completely extinguish the X-ray and is, therefore, responsible for well-known line-artifacts in computed tomograms. Artifacts are caused by abrupt transitions between low- and high-density materials, which results in data values that exceed the dynamic range of the processing electronics.

Contrast mediums used for X-ray CT, as well as for plain film X-ray, are called radiocontrasts. Radiocontrasts for X-ray CT are, in general, iodine-based. Often, images are taken both with and without radiocontrast. CT images are called precontrast or native-phase images before any radiocontrast has been administrated, and postcontrast after radiocontrast administration.

Two-dimensional CT images are conventionally rendered so that the view is as though looking up at it from the patient's feet. Hence, the left side of the image is to the patient's right and vice versa, while anterior in the image also is the patient's anterior and vice versa. This left-right interchange corresponds to the view that physicians generally have in reality when positioned in front of patients.

CT data sets have a very high dynamic range which must be reduced for display or printing. This is typically done via a process of "windowing", which maps a range (the "window") of pixel values to a greyscale ramp. For example,

CT images of the brain are commonly viewed with a window extending from 0 HU to 80 HU. Pixel values of 0 and lower, are displayed as black; values of 80 and higher are displayed as white; values within the window are displayed as a grey intensity proportional to position within the window. The window used for display must be matched to the X-ray density of the object of interest, in order to optimize the visible detail.

THREE-DIMENSIONAL RECONSTRUCTION

Because contemporary CT scanners offer isotropic or near isotropic, resolution, display of images does not need to be restricted to the conventional axial images. Instead, it is possible for a software program to build a volume by "stacking" the individual slices one on top of the other. The program may then display the volume in an alternative manner.

Multiplanar reconstruction

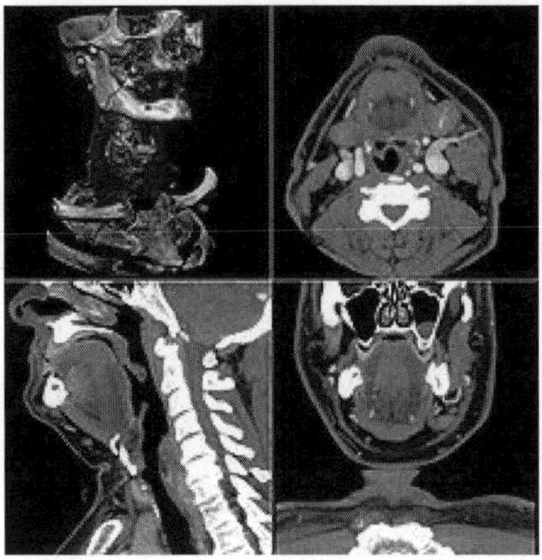

Fig. Typical screen layout for diagnostic software, showing one 3D and three MPR views

Multiplanar reconstruction (MPR) is the simplest method of reconstruction. A volume is built by stacking the axial slices. The software then cuts slices through the volume in a different plane (usually orthogonal). As an option, a special projection method, such as maximum-intensity projection (MIP) or minimum-intensity projection (mIP/MinIP), can be used to build the reconstructed slices.

MPR is frequently used for examining the spine. Axial images through the spine will only show one vertebral body at a time and cannot reliably show the intervertebral discs. By reformatting the volume, it becomes much easier to visualise the position of one vertebral body in relation to the others.

Modern software allows reconstruction in non-orthogonal (oblique) planes so that the optimal plane can be chosen to display an anatomical structure. This may be particularly useful for visualising the structure of the bronchi as these do not lie orthogonal to the direction of the scan. For vascular imaging, curved-plane reconstruction can be performed. This allows bends in a vessel to be "straightened" so that the entire length can be visualised on one image, or a short series of images. Once a vessel has been "straightened" in this way, quantitative measurements of length and cross sectional area can be made, so that surgery or interventional treatment can be planned.

MIP reconstructions enhance areas of high radiodensity, and so are useful for angiographic studies. MIP reconstructions tend to enhance air spaces so are useful for assessing lung structure.

3D rendering techniques

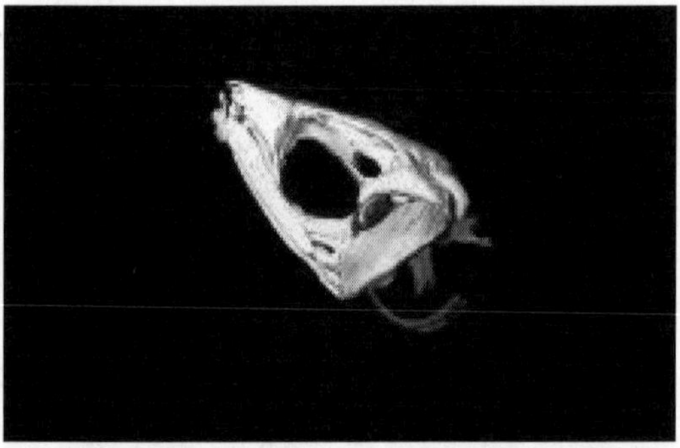

Fig. Surface rendering of the head of the frog Atelopus franciscus, with ear parts highlighted.

Surface rendering

A threshold value of radiodensity is set by the operator (e.g., a level that corresponds to bone). From this, a three-dimensional model can be constructed using edge detection image processing algorithms and displayed on screen.

Multiple models can be constructed from various thresholds, allowing different colors to represent each anatomical component such as bone, muscle, and cartilage. However, the interior structure of each element is not visible in this mode of operation.

Volume rendering

Surface rendering is limited in that it will display only surfaces that meet a threshold density, and will display only the surface that is closest to the imaginary viewer. In volume rendering, transparency, colors and shading are used to allow a better representation of the volume to be shown in a single image. For example, the bones of the pelvis could be displayed as semi-transparent, so that, even at an oblique angle, one part of the image does not conceal another.

Image segmentation

Where different structures have similar radiodensity, it can become impossible to separate them simply by adjusting volume rendering parameters. The solution is called segmentation, a manual or automatic procedure that can remove the unwanted structures from the image.

IMAGE QUALITY

Artifacts

Although images produced by CT are generally faithful representations of the scanned volume, the technique is susceptible to a number of artifacts, such as the following:

Streak artifact

Streaks are often seen around materials that block most X-rays, such as metal or bone. Numerous factors contribute to these streaks: undersampling, photon starvation, motion, beam hardening, and Compton scatter. This type of artifact commonly occurs in the posterior fossa of the brain, or if there are metal implants. The streaks can be reduced using newer reconstruction techniques or approaches such as metal artifact reduction (MAR).

Partial volume effect

This appears as "blurring" of edges. It is due to the scanner being unable to differentiate between a small amount of high-density material (e.g., bone) and a larger amount of lower density (e.g., cartilage). The reconstruction assumes that the X-ray attenuation within each voxel is homogenous; this may not be the case at sharp edges. This is most commonly seen in the z-direction, due to the conventional use of highly anisotropic voxels, which have a much lower out-of-plane resolution, than in-plane resolution. This can be partially overcome by scanning using thinner slices, or an isotropic acquisition on a modern scanner.

Ring artifact

Probably the most common mechanical artifact, the image of one or many "rings" appears within an image. This is usually due to a detector fault, or miscalibration of an individual detector element.

Noise

This appears as grain on the image and is caused by a low signal to noise ratio. This occurs more commonly when a thin slice thickness is used. It can also occur when the power supplied to the X-ray tube is insufficient to penetrate the anatomy.

Motion artifact

This is seen as blurring and/or streaking, which is caused by movement of the object being imaged. Motion blurring might be reduced using a new technique called IFT (incompressible flow tomography).

Windmill

Streaking appearances can occur when the detectors intersect the reconstruction plane. This can be reduced with filters or a reduction in pitch.

Beam hardening

This can give a "cupped appearance". It occurs when there is more attenuation along a path passing through the center of an object, than a path that grazes the edge. This is easily corrected by filtration and software.

Dose vs. image quality

An important issue within radiology today is how to reduce the radiation dose during CT examinations without compromising the image quality. In general, higher radiation doses result in higher-resolution images, while lower doses lead to increased image noise and unsharp images.

However, increased dosage raises increase the adverse side effects, including the risk of radiation induced cancer – a four-phase abdominal CT gives the same radiation dose as 300 chest x-rays (See the Scan dose section). Several methods that can reduce the exposure to ionizing radiation during a CT scan exist.

1. New software technology can significantly reduce the required radiation dose.
2. Individualize the examination and adjust the radiation dose to the body type and body organ examined. Different body types and organs require different amounts of radiation.
3. Prior to every CT examination, evaluate the appropriateness of the exam whether it is motivated or if another type of examination is more suitable. Higher resolution is not always suitable for any given scenario, such as detection of small pulmonary masses.

INDUSTRIAL USE

Industrial CT Scanning (industrial computed tomography) is a process which utilizes x-ray equipment to produce 3D representations of components both externally and internally. Industrial CT scanning has been utilized in many areas of industry for internal inspection of components. Some of the key uses for CT scanning have been flaw detection, failure analysis, metrology, assembly analysis, and reverse engineering applications. CT scanning is also employed in the imaging and conservation of museum artifacts.

HISTORY

Origins of tomography

In the early 1900s, the Italian radiologist Alessandro Vallebona proposed a method to represent a single slice of the body on the radiographic film. This method was known as tomography. The idea is based on simple principles of projective geometry: moving synchronously and in opposite directions the X-ray tube and the film, which are connected together by a rod whose pivot point is the focus; the image created by the points on the focal plane appears sharper, while the images of the other points annihilate as noise. This is only marginally effective, as blurring occurs in only the "x" plane. There are also more complex devices that can move in more than one plane and perform more effective blurring.

Mathematical theory

The mathematical theory behind the tomographic reconstruction dates back to 1917 where the invention of Radon Transform by an Austrian mathematician Johann Radon. He showed mathematically that a function could be reconstructed from an infinite set of its projections. In 1937, a Polish mathematician, named Stefan Kaczmarz, developed a method to find an approximate solution to a large system of linear algebraic equations. This led the foundation to another powerful reconstruction method called "Algebraic Reconstruction Technique (ART)" which was later adapted by Sir Godfrey Hounsfield as the image reconstruction mechanism in his famous invention, the first commercial CT scanner.

In 1956, Ronald N. Bracewell used a method similar to the Radon Transform to reconstruct a map of solar radiation from a set of solar radiation measurements. In 1959, William Oldendorf, a UCLA neurologist and senior medical investigator at the West Los Angeles Veterans Administration hospital, conceived an idea for "scanning a head through a transmitted beam of X-rays, and being able to reconstruct the radiodensity patterns of a plane through the head" after watching an automated apparatus built to reject frostbitten fruit by detecting dehydrated portions.

In 1961, he built a prototype in which an X-ray source and a mechanically coupled detector rotated around the object to be imaged. By reconstructing the image, this instrument could get an X-ray picture of a nail surrounded by a circle of other nails, which made it impossible to X-ray from any single angle. In his landmark paper published in 1961, he described the basic concept which was later used by Allan McLeod Cormack to develop the mathematics behind computerized tomography.

In October 1963, Oldendorf received a U.S. patent for a "radiant energy apparatus for investigating selected areas of interior objects obscured by dense material." Oldendorf shared the 1975 Lasker award with Hounsfield for that discovery. The field of the mathematical methods of computerized tomography has seen a very active development since then, as is evident from overview literature by Frank Natterer and Gabor T. Herman, two of the pioneers in the this field.

Tomography has been one of the pillars of radiologic diagnostics until the late 1970s, when the availability of minicomputers and of the transverse axial scanning method gradually supplanted it as the modality of CT. Transverse axial scanning was due in large part to the work of Godfrey Hounsfield and South African-born Allan McLeod Cormack. In terms of mathematics, the method is based upon the use of the Radon Transform. But as Cormack remembered later, he had to find the solution himself since it was only in 1972 that he learned of the work of Radon, by chance.

Commercial scanners

The first commercially viable CT scanner was invented by Sir Godfrey Hounsfield in Hayes, United Kingdom, at EMI Central Research Laboratories using X-rays. Hounsfield conceived his idea in 1967. The first EMI-Scanner was installed in Atkinson Morley Hospital in Wimbledon, England, and the first patient brain-scan was done on 1 October 1971. It was publicly announced in 1972.

The original 1971 prototype took 160 parallel readings through 180 angles, each 1° apart, with each scan taking a little over 5 minutes. The images from these scans took 2.5 hours to be processed by algebraic reconstruction techniques on a large computer. The scanner had a single photomultiplier detector, and operated on the Translate/Rotate principle.

The success of The Beatles enabled EMI to fund research and build early models for medical use. The first production X-ray CT machine (in fact called the "EMI-Scanner") was limited to making tomographic sections of the brain, but acquired the image data in about 4 minutes (scanning two adjacent slices), and the computation time (using aData General Nova minicomputer) was about 7 minutes per picture.

This scanner required the use of a water-filled Perspex tank with a pre-shaped rubber "head-cap" at the front, which enclosed the patient's head. The water-tank was used to reduce the dynamic range of the radiation reaching the detectors (between scanning outside the head compared with scanning through the bone of the skull). The images were relatively low resolution, being composed of a matrix of only 80 x 80 pixels.

In the U.S., the first installation was at the Mayo Clinic. As a tribute to the impact of this system on medical imaging the Mayo Clinic has an EMI scanner on display in the Radiology Department. Allan McLeod Cormack of Tufts University in Massachusetts independently invented a similar process, and both Hounsfield and Cormack shared the 1979 Nobel Prize in Medicine.

The first CT system that could make images of any part of the body and did not require the "water tank" was the ACTA (Automatic Computerized

Transverse Axial) scanner designed by Robert S. Ledley, DDS, at Georgetown University. This machine had 30 photomultiplier tubes as detectors and completed a scan in only 9 translate/rotate cycles, much faster than the EMI-scanner. It used a DEC PDP11/34 minicomputer both to operate the servo-mechanisms and to acquire and process the images. The Pfizer drug company acquired the prototype from the university, along with rights to manufacture it. Pfizer then began making copies of the prototype, calling it the "200FS" (FS meaning Fast Scan), which were selling as fast as they could make them. This unit produced images in a 256×256 matrix, with much better definition than the EMI-Scanner's 80×80.

Since the first CT scanner, CT technology has vastly improved. Improvements in speed, slice count, and image quality have been the major focus primarily for cardiac imaging. Scanners now produce images much faster and with higher resolution enabling doctors to diagnose patients more accurately and perform medical procedures with greater precision. In the late 90's CT scanners broke into two major groups, "Fixed CT" and "Portable CT".

"Fixed CT Scanners" are large, require a dedicated power supply, electrical closet, HVAC system, a separate workstation room, and a large lead lined room. "Fixed CT Scanners" can also be mounted inside large tractor trailers and driven from site to site and are known as "Mobile CT Scanners". "Portable CT Scanners" are light weight, small, and mounted on wheels. These scanners often have built-in lead shielding and run off of batteries or standard wall power.

Etymology

The word "tomography" is derived from the Greek tomos (slice) and graphein (to write). Computed tomography was originally known as the "EMI scan" as it was developed in the early 1970s at a research branch of EMI, a company best known today for its music and recording business. It was later known as computed axial tomography (CAT or CT scan) and body section röntgenography.

Although the term "computed tomography" could be used to describe positron emission tomography or single photon emission computed tomography (SPECT), in practice it usually refers to the computation of tomography from X-ray images, especially in older medical literature and smaller medical facilities.

In MeSH, "computed axial tomography" was used from 1977 to 1979, but the current indexing explicitly includes "X-ray" in the title.

TYPES OF MACHINES

Spinning tube, commonly called spiral CT, or helical CT in which an entire X-ray tube is spun around the central axis of the area being scanned. These are the dominant type of scanners on the market because they have been manufactured longer and offer lower cost of production and purchase. The main limitation of this type is the bulk and inertia of the equipment (X-ray tube assembly and detector array on the opposite side of the circle) which limits the speed at which the equipment can spin. Some designs use two X-ray sources and detector arrays offset by an angle, as technique to improve temporal resolution.

Electron beam tomography (EBT) is a specific form of CT in which a large enough X-ray tube is constructed so that only the path of the electrons, travelling between the cathode and anode of the X-ray tube, are spun using deflection coils. This type had a major advantage since sweep speeds can be much faster, allowing for less blurry imaging of moving structures, such as the heart and arteries. Fewer scanners of this design have been produced when compared with spinning tube types, mainly due to the higher cost associated with building a much larger X-ray tube and detector array and limited anatomical coverage. Only one manufacturer (Imatron, later acquired by General electric) ever produced scanners of this design. Production ceased in early 2006.

Previous studies

Pneumoencephalography of the brain was quickly replaced by CT. A form of tomography can be performed by moving the X-ray source and detector during an exposure. Anatomy at the target level remains sharp, while structures at different levels are blurred. By varying the extent and path of motion, a variety of effects can be obtained, with variable depth of field and different degrees of blurring of "out of plane" structures. Although largely obsolete, conventional tomography is still used in specific situations such as dental imaging (orthopantomography) or in intravenous urography.

4

Forensic DNA Analysis

DNA FINGERPRINTING

DNA analysis or 'DNA fingerprinting' was first described in 1985 by Dr. Alec Jeffreys, an English geneticist. Jeffreys discovered that DNA contained a number of sequences that were repeated again and again. He then established that the number of repeated sections varied in each individual, rendering their genetic make-up completely unique, with the exception of identical twins. Jeffreys developed a technique which could examine the variation of length in each of these DNA sequences which created the ability to discern one person from another. The concept of testing DNA for the purpose of human identification was then established.

DNA can be extracted from a variety of places such as blood, semen, bones and teeth. DNA analysis does not mean testing every single base of the DNA as this would be an impossible feat. Instead scientists test a location in the DNA known as a locus. The number of loci which will be tested varies from jurisdiction to jurisdiction.

Since its first use in 1985, DNA analysis has developed scientifically following the introduction of a number of sensitive and accurate scientific tools and techniques. One such technique is Low Copy Number (LCN) DNA which has led to attempts to analyse more difficult and challenging samples such as those containing DNA from only a few cells. Today, the sensitivity and discriminating power of forensic DNA analysis has resulted in the science behind this important investigative technique becoming even more sensitive (scientifically) and the public perception of it even more powerful.

THE ADVANTAGES OF DNA PROFILING

Forensic DNA analysis has many advantages. Firstly, DNA evidence can be said to be a more reliable form of evidence which is now widely accepted by the scientific community. Prior to the use of DNA evidence in courts, eye witness testimony was more heavily relied upon and this brought along with it inherent dangers. The Mid-Atlantic Innocence Project revealed that mistaken eyewitness identifications were a factor in more than 70% of the initial 239

DNA exoneration cases. This is a huge percentage which highlights the difficulties with relying on eyewitness identification as a reliable source of evidence. DNA analysis sought to relieve some of the problems which existed in the criminal justice system at that time. An authoritative study on the forensic use of DNA noted that; "...the reliability of DNA evidence will permit it to exonerate some people who would have been wrongfully accused or convicted without it."

Secondly, DNA analysis can also be seen to be valid as it provides a scientific basis which allows for a physical link to be made between a criminal and a crime scene in order to secure a conviction. At the same time it also works to exclude suspects who without it may be charged for a crime they did not commit. The validity of DNA analysis is particularly clear in cases where DNA has been used to exonerate those wrongly convicted.

THE FIRST DNA EXONERATION CASE

Forensic use of DNA technology was first used to exclude a suspect in 1986 in the English case of Colin Pitchfork which arose when two young girls were raped and murdered in Leicestershire, in 1983 and 1986. The first murder was that of a 15 year old school girl, Lynda Mann. Blood discovered at the scene was found to be blood type A (which at that time amounted to a 10% match of the adult male population). Due to no further leads and a lack of forensic evidence the police had no other option but to wind down the investigation into the murder.

Three years later in 1986 however, another 15 year old girl, Dawn Ashworth, was found raped and murdered in the same town. The police were convinced that both murders had been committed by the same person. Semen samples from the second murder also confirmed a match to the blood type of the first. A local man then confessed to the murder of Dawn Ashworth, yet denied any involvement in the first murder. Police consulted Jeffreys, who (as discussed above) had developed a technique that could examine DNA profiles, in an attempt to verify that the suspect was responsible for both of the murders.

In 1985, Jeffreys along with Dr Peter Gill and Dr Dave Werrett had been the first to demonstrate that DNA could be lifted from stains left at a crime scene, a point which proved vital in the case of Colin Pitchfork. DNA tests were conducted and were able to establish that the suspect was not responsible for the murders. The police then conducted a mass operation to obtain blood samples from 4,000 men in the area. Initially, no matches were found, however, it was then discovered that Pitchfork had made his friend give DNA on his behalf. His friend was later overheard discussing this and Colin Pitchfork was arrested. His DNA was then found to be a match to the crime scenes. This case was the first in the world to exonerate a suspect through the use of DNA evidence. Jeffreys later said "I have no doubt whatsoever that he [the man who had falsely confessed] would have been found guilty had it not been for DNA evidence. That was a remarkable occurrence." Had it not been for the introduction of forensic DNA profiling the real criminal, Colin Pitchfork, may never have been found and an innocent person could have been wrongfully imprisoned.

INNOCENCE PROJECTS

Forensic DNA profiling, many non-profit legal organisations known as Innocence Projects were set up, dedicated to using the new technique to help exonerate those who had been wrongly convicted. Since their introduction Innocence Projects have had a significant impact using DNA profiling to rectify miscarriages of justice around the world. In the US alone there have been 275 post-conviction DNA exoneration cases since the first in 1989. 208 of these exoneration cases have been since the year 2000.

Out of the 275 exoneration cases, 17 of the people served time on death row. Had it not been for DNA proving their innocence they may not be alive today. The average prison sentence served by exonerates is 13 years and in total the number of years served over all exoneration cases is approximately 3,564. The impact of Innocence Projects in utilising forensic DNA profiling in rectifying miscarriages of justice speaks for itself upon reading the above statistics.

Yet another advantage of DNA analysis is that it is objective, in that the results are completely factual. Unlike eyewitness testimony, there are no issues of personal feelings or opinions involved in obtaining a result; instead it is based on scientific processes which are able to produce profiles which can then be interpreted to determine the likelihood of a match between a DNA sample found at the crime scene and the DNA of a suspect.

CONCLUSION

DNA analysis has proved itself time and again to be an extremely powerful scientific tool which carries with it a great deal of general acceptance by both the scientific and legal communities. At first instance forensic DNA profiling appears to be reliable, valid and objective. This raises questions however: Just how heavy or simplistic a reliance should there be on the results of DNA analysis, when so much is at stake depending on the interpretation of the results given by the forensic scientist?

Is there a perception that DNA is almost infallible and conclusive in all respects? Should there be a greater awareness of the dangers of over reliance or over-simplification of the interpretation of DNA results, which could almost prove as dangerous and as unlawful as a misidentification by an eyewitness? The remainder of this thesis will be dedicated to addressing these issues in order to examine the dangers of an over reliance and over-simplification of forensic DNA profiling results.

THE RELIABILITY & ADMISSIBILITY OF DNA EVIDENCE

The remarkable success of forensic DNA profiling has led to attempts to analyse more difficult and challenging samples such as those containing DNA from only a few cells. This approach, known as Low Copy Number (LCN) DNA typing; 'facilitates the examination of a whole new range of evidence types that previously could not be analysed because of the very low amounts of DNA recoverable from the sample'. In general, LCN DNA testing refers to testing a sample which contains less than 100pg of DNA. Laboratories employ a number of techniques to do so such as increasing the number of Polymerase Chain Reaction (PCR) cycles to improve the amplification field from samples containing low levels of DNA. With increased cycles of PCR, samples that were originally very small can then be copied so many times that they become able to be analysed. The impact of LCN on the criminal justice system is significant and likely only to increase in the future.

In recent years, a number of high profile cases have allowed the courts to express their opinion as to the reliability, admissibility and evidential value of DNA evidence obtained using the LCN process. The validity of LCN DNA has been a controversial matter at least since R v Hoey in 2007 and more recently in R v Reed. This chapter discusses the special considerations which are required to interpret the results of LCN DNA given that it is impossible to tell where such a minute sample came from. In particular, it is important to consider the implications of allele dropout and the possibility of laboratory-based contamination. The remainder of this chapter will consider the legal admissibility of expert opinion evidence and what is required for it to be deemed admissible by the courts.

THE RISKS OF LCN DNA

The sensitive nature of the LCN DNA process is accompanied by a range of risks which may lead to possible wrongful convictions and may also have the potential to mislead criminal investigations. One such risk comes from the number of PCR cycles which have to be considerably increased to obtain an LCN DNA profile. PCR is a common technique to amplify a number of copies of a piece of DNA generating thousands to millions of a particular DNA sequence. Increasing PCR cycles inevitably leads to a magnified risk of contamination and inaccurate results caused from 'stochastic effects'. Stochastic effects occur predominantly when only a very small amount of DNA is available to begin with, and materialise when random loci or alleles are sampled more than others, leading to peak height imbalance and causing alleles to drop out completely. The trouble is that it is difficult to tell which if any peaks are missing or falsely present, and even if the sample was to be run through the machine a number of times, the same result may not necessarily be produced each time - which can clearly lead to unreliable test results. This is a major problem with LCN, because the cornerstone of good scientific method requires that results are reproducible. What this means for a criminal case is that it may produce a result that is a whole or partial profile that does or does not match the accused, but if the results cannot be produced reliably – how reliable is it all?

Another risk accompanied with the LCN process is contamination. If the starting amount of DNA is very small and there is also some contamination in it, if PCR is used to multiply the sample – the contaminant is also multiplied. As the amount of crime sample DNA decreases, the chance of contamination by other sources increases; the DNA contamination will then be multiplied along with the suspects DNA. This then becomes difficult (and sometimes) impossible to determine what DNA is from the offender and what is contamination. Whereas if you had a large amount of DNA to begin with, big peaks are evident in the DNA results and tiny contaminant peaks are conspicuous and easily identifiable. The trouble with LCN is, all the peaks are tiny and it becomes difficult to tell (scientifically) which are contaminants and which are real.

LCN DNA RESULTS: THE DIFFICULTY OF INTERPRETATION

As a result of above risks, there are a number of difficulties associated with the interpretations that can be drawn from LCN DNA results even if a DNA profile is accurately yielded. A phenomenon termed 'adventitious transference' can occur due to the fact that an LCN profile can stem from the cells of a single touch which may have originated from innocent interactions by individuals unrelated to the crime. This is another significant pitfall due to the advances in technology that make it possible to detect and test increasingly minute DNA samples. Thomson et al. (2003) highlight, "Whereas the original DNA tests required a fairly large amount of biological material to get a result (e.g. a blood stain the size of a dime), current DNA tests are so sensitive that they can type the DNA found in samples containing only a few cells."

Although adventitious transfer cannot be strictly controlled, there are a number of ways to minimise and account for such contamination including: improve sample collection and if possible obtain a DNA sample from the normal user of an exhibit of interest or the person who may have come into contact with the exhibit prior to the offence. This is very difficult in major crime scenes or scenes which are busy public places as there are a number of ways to contaminate a crime scene. Contamination can take place if someone sneezes, coughs or touches a part of their body and then touches the area containing the sample to be tested. DNA left at a crime scene can also be subject to environmental contamination such as exposure to heat, light and moisture which can speed up degradation of DNA. As a result, not all DNA evidence may yield a usable profile.

Overall, Gill (2001) noted three specific consequences of amplifying LCN DNA which can lead to inaccurate DNA profiles; (a) allele drop-out may occur because one allele can be preferentially amplified; (b) stutters in the profile may be preferentially analysed – sometimes referred to as false alleles; and (c) the method is prone to sporadic contamination which occurs when alleles that are not associated with the crime stain or sample are amplified. This could lead to complex or incomplete profiles, and inaccurate interpretation of these profiles could lead to wrongful convictions. Even worse, the extreme sensitivity of LCN increases the possibility of false forensic inferences.

LCN DNA: PUBLIC CONCERNS

The scientific controversy surrounding LCN DNA analysis rose to public attention in the case of Sean Hoey. In August 1998, a car bomb exploded in a busy shopping centre in the Northern Ireland town of Omagh, killing 29 people and injuring over 200 others. The bombing was described by the media as "the worst terrorist atrocity in Northern Ireland's history". A key part of the trial involved LCN DNA analysis by the Forensic Science Service (FSS) on pieces of the bomb timer and explosive wiring. Court Records state: "The evidence establishes that the arrangements within the police in 1998 and 1999 for the recording and storage of items were thoroughly disorganised." The property store was described as "a complete mess." There was no universal system of logging items received, no proper recording in police stations so no inventory of what was in a store room at a particular time – all of which significantly increased the risk of contamination.

The trial lasted 56 days and was hinged on LCN DNA evidence on the basis of an apparent LCN DNA link between Hoey and a number of exhibits recovered from crime scenes. When the trial concluded in December 2007, Mr Hoey was found not guilty of all 58 charges brought against him. Even more significantly the judge - Justice Weir - strongly rejected the evidence, raising significant concerns about the validity of the LCN technique and also highlighting the careless handling of DNA throughout the trial. Fundamentally, the DNA evidence was found not to have the necessary integrity to be reliable due to the risk of it having been contaminated or interfered with. In his concluding observation Weir J emphasised the cardinal principle of criminal law established in R v Steenson; "Justice 'according to law' demands proper evidence… "evidence which is so convincing in truth and manifestly reliable that it reaches the standard of proof beyond reasonable doubt." The evidence in Hoey failed to meet such an immutable standard and accordingly Mr Hoey was found not guilty.

This judgement raised considerable questions over the merits of LCN DNA testing and led to a review of the technique. An interim suspension was placed on the use of LCN DNA in criminal investigations in England and Wales while an internal review of on-going cases involving the FSS and the LCN DNA technique was conducted by the Crown Prosecution Service (CPS). The CPS concluded "that LCN DNA analysis provided by the FSS should remain available as potentially admissible evidence" The press release went on to state that "At present, there is no reason to believe that there is any inherent unreliability in the LCN DNA analysis process provided that it is carried out according to the prescribed processes, and that the results are properly interpreted. In its work so far, the review has found nothing that would indicate any serious flaw in the scientific principles." The nature of the samples makes reproducibility a significant problem for obtaining a consistent profile and for giving a reasoned interpretation of what the results mean.

Following the judgement in Hoey, a review (known as the 'The Caddy Review') was conducted into the low template DNA profiling techniques. The authors of the Caddy Review state that "it is our opinion that LCN and LTD [NA] are extensions of the internationally accepted process of standard DNA profiling". The authors further concluded that "reservations (with regard to the LCN technique) have been allayed from a study of the raw data produced by the FSS, recent experimental work conducted by the FSS and also from detail information submitted by the other (UK) forensic science providers which clearly demonstrate the soundness of LTDNA analysis (including LCN) providing all the appropriate conditions are met" Emphasis was once again placed on the importance of satisfying all the appropriate conditions and ensuring that the LCN DNA analysis is carried out according to the prescribed processes.

The Caddy Review also chose to emphasise that "any LTDNA profile should always be reported to the jury with the caveats: that the nature of the original starting material is unknown that; the time at which the DNA was transferred cannot be inferred; and that the opportunity for secondary transfer is increased in comparison to standard DNA profiling". The juror's interpretation of DNA evidence is an issue which will be discussed in more depth in subsequent chapters however; it is important to note the emphasis placed on presenting an LTDNA profile to the jury in comparison to a standard DNA profile.

LTN DNA results must be interpreted cautiously and all factors regarding the extreme sensitivity of the technique must be taken into consideration. The Caddy Review gave the LCN DNA technique an apparent clean bill of health however the conclusions set out in the review have since attracted a great deal of criticism and it would appear that there still exists a great deal of ambiguity with regards to the general uses of the LCN DNA technique. There is a significant risk that the Crown can read more into results of LCN than they should (i.e. to inculpate an accused person), given that they cannot specifically determine the source of the sample, the extent of the contamination and transfer.

The English Court of Appeal has expressed its opinion as to the reliability, admissibility and evidential value of low template DNA primarily that obtained using the LCN process in the case of R v Reed. The judgement in this case deals with two main issues. The first is the admissibility of LCN DNA as evidence. The second is whether an expert witness should be permitted to provide an opinion on the LCN DNA evidence on matters such as the likelihood and means by which the DNA may have been transferred to the place of discovery.

Overall, the judgement concluded that: "... a challenge to the validity of the method of analysing Low Template DNA by the LCN process should no longer be permitted at trials where the quantity of DNA analysed is above the stochastic threshold of 100–200 picograms [with a picogram being one trillionth of a gram] in the absence of new scientific evidence." The judgement thus sets a minimum standard for LCN DNA evidence as to what is required for the court to deem it as a reliable form of scientific evidence.

The Legal Admissibility of Dna Evidence

Given everything discussed about the scientific reliability of LCN, it is now imperative to discuss the legal admissibility of this technique. The admissibility of scientific evidence is an issue which has plagued the courts for a number of years, and one which The Court of Appeals of the District of Columbia addressed in Frye v United States. The Court in Frye stated that "Just when a scientific principle or discovery crosses the line between the experimental and demonstrable stages is difficult to define... the thing from which the deduction is made must be sufficiently established to have gained general acceptance in the particular field in which it belongs." The Frye test stated that any new evidence should be approved by the people who use it, however many questions remained unanswered after Frye: What types of evidence shall be accepted?

If accepted, what shall the purpose of it be? Who shall accept it? How many experts constitute general acceptance? Despite the ambiguities in the exact meaning of Frye, it was utilised by many courts in the 70 years after, in determining the admissibility of novel scientific evidence. The first case to question Frye and the idea that general acceptance in itself was sufficient for scientific evidence to be admissible, arose in Daubert v Merrell Dow. In this case, the US Supreme Court ruled that the trial judge was the "gatekeeper" to prevent "junk science" from entering the courtroom. This gate keeping function of the court could be interpreted to suggest that the court will be wary of sciences in order to prevent evidence with no scientific basis from even entering the court.

The US Supreme court in Daubert recognised "general acceptance" as an important, rather than exclusive, factor in determining the reliability of scientific evidence. The implication of this would be a higher standard of test for determining what types of evidence are scientific which may be another step in the right direction to prevent so-called "junk science" from entering the court room. The court suggested that there are four factors to consider, now known as the "Daubert factors", they are; testing and validation, peer review, error rates, and lastly, the traditional factor set out in Frye of general acceptance in the relevant scientific community. In order to be deemed admissible, scientific evidence would need to satisfy all of the above criteria. Daubert also placed increased emphasis on the overall reliability of scientific evidence.

In April 2009, the Law Commission published Consultation Paper 190, 'The Admissibility of Expert Evidence in Criminal Proceedings in England and Wales: A New Approach to the Determination of Evidentiary Reliability'. The Consultation was elicited by a growing dissatisfaction with the state of the law relating to the admission of expert evidence. The Commission were concerned that there exists a "pressing danger" of wrongful convictions and acquittals and that the most recent miscarriages of justice may be 'the tip of a larger iceberg'. The Commission identified their proposal as the progeny of Daubert. The Commission, however, unlike Daubert, proposed two sets of guidelines: one set of guidelines relate to scientific evidence and the second to experience based testimony.

The Commission proposed that expert evidence would be admissible in criminal proceedings only if certain tests were met: firstly the court must be satisfied that it would provide information which is likely to be outside a judge or jury's experience or knowledge, and which would give them help they need in arriving at their conclusions; secondly the witnesses must be qualified; thirdly the evidence is not made inadmissible as a result of clause 3 (Impartiality). The Commission also proposed a new statutory provision that expert opinion evidence is admissible only if the court is satisfied that it is sufficiently reliable to be admitted. The distinctive feature of this proposal lay in the Commission's clear articulation of the methodology required to establish reliability.

The Commission concluded that the present law common law approach to the admissibility of expert evidence in criminal proceedings in England and Wales is unsatisfactory and should be replaced with a new statutory test. The Commission suggested that a Daubert-style reliability test in tandem with appropriate training for the judiciary and practitioners, offers the best mechanism for excluding unreliable expert evidence. The Commission set out four proposals aimed at ensuring expert evidence will be submitted only if it is sufficiently reliable and ultimately acted out upon by the jury.

The key proposal was that there should be an explicit "gate-keeping" role for the trial judge with a clearly-defined test for determining whether proffered expert evidence is sufficiently reliable to be admitted. The Commission also proposed that there should exist the onus of persuasion - if a party proposes to rely on expert evidence, any party against whom the evidence would be adduced should be entitled to raise the question of its evidentiary reliability. In addition the Committee also noted that the judge should be permitted to call upon an independent assessor to provide him or her with assistance and guidance in cases where the evidence of field is particularly difficult. The Commission also noted that judges (and criminal practitioners) should receive practical training on the methodology of science, the standards for determining the statistical significance of research findings and how to determine the reliability of experience-based expertise.

The Law Commission's proposal on expert opinion evidence has since been criticised, with one article claiming that whilst the criteria set out in the paper are clearly articulated applying them to forensic disciplines is 'fraught with difficulty.' The article also suggests that the proposal places an onerous demand upon judges, who have the difficulty of interpreting the criteria. If interpreted too strictly it could lead to the exclusion of forensic testimony becoming widespread. On the other hand, if interpreted too liberally, the criteria will not provide adequate safeguards against admission of unfounded evidence.The article rejects the criteria based approach towards admission, suggesting instead, that forensic science should be evaluated by carefully constituted working parties outside the courtroom.

DNA profiling technology has improved greatly since the days when a visible blood drop or reasonable sized stain was required to obtain a profile. New techniques such as LCN now allow for the analysis of more difficult and challenging samples which may contain DNA from only a few cells. These new techniques are undoubtedly of tremendous value in criminal investigations, however, failure to interpret DNA profiles using standard scientific principles will almost certainly result in significant damage, although unintentional, that could threaten it's more restricted but beneficial use. The sensitive nature of the LCN technique is accompanied by a number of risks which can lead to the results failing to be reproducible or accurate. If results cannot be reproduced reliably or it becomes impossible to scientifically determine the source of the DNA then how can the results of this technique be reliable?

The risks of the LCN technique also mean that there are a number of difficulties associated with the interpretation of the results even if a DNA profile is accurately yielded. The consequences of amplifying LCN DNA may lead to complex or incomplete profiles which coupled with inaccurate interpretation could then lead to wrongful convictions. Despite the inherent dangers of such a technique reviews have concluded that there is no reason to believe there is any "inherent unreliability" in the LCN analysis if it is carried out in accordance with prescribed processes and if the results are properly interpreted. The onus is then on the judge to scrupulously protect the jury from unreliable scientific evidence.

The following chapters intend to look at the dangers which exist if the results of forensic DNA analysis are incorrectly interpreted and just how essential it is to correctly analyse the reliability, validity and accuracy of the results to prevent future miscarriages of justice. In doing so, it will look at the difficulties facing the forensic science community who must scientifically interpret the results and the jury who must reach a verdict based on the results.

THE SCIENTIFIC INTERPRETATION OF DNA EVIDENCE

Scientific evidence can be problematic for courts: the subject matter is often complex, the experts can be biased and the judge and juries may be unsure of how to interpret the evidence once it has been presented in court. At an even more fundamental level, miscarriage of justice will arise if the science underlying the expert evidence presented in Court is not sound. Interpretation of forensic evidence often has to be performed within a framework appropriate for both scientific analysis and presentation in court.

Forensic DNA analysts often testify to their findings using statistical probabilities rather than statements of certainty which can turn out to be arbitrary and unsupported. The field of DNA analysis still requires a great deal of analysis as the apparent certainty of this powerful investigative tool can be deceptive and this can lead to it being misused and misapplied. There are many ways to introduce results to the jury and it is important for DNA results and results from a number of other disciplines to be accurately presented to the jury. One discipline which will be discussed is the field of shoe mark analysis.

DIFFERENT MODES OF PRESENTING RESULTS

In the case of R v T, shoe marks were recovered from the crime scene by the forensic footwear expert, Mr Ryder, which he compared to shoes which had been recovered from the defendant's property. Mr Ryder concluded that "... it is somewhat unlikely that the observed correspondence would have been obtained as a result of mere coincidence had the recovered footwear not made the mark in question... there is at this stage a moderate degree of scientific evidence to support the view that the shoes recovered had made the footwear marks."

The question before the court was therefore whether the Bayesian approach of using likelihood ratios had been applied in this case, an approach which the court had '...robustly rejected for non DNA evidence in a number of cases'. The Court of Appeal allowed T's appeal against conviction because it became apparent that 'the Bayesian approach" of using likelihood ratios' had been applied and this 'had not been explored in the course of the trial.

Many leading scientists are of the opinion that the likelihood-ratio framework is the most logically correct framework for the evaluation of forensic science evidence and should be applied irrespective of the structure of the data extracted from the objects. The idea that the likelihood-ratio framework is the most logical approach to evaluating forensic science evidence dates back to the early 1900s and it became standard practice in forensic DNA comparison in the mid-1990's. It has since been adopted by the Forensic Science Service (FSS) and the Netherlands Forensic Institute.

Throughout R v T the Court of Appeal heard from a number of expert witnesses, including the principal scientist at the Forensic Science Service (FSS) responsible for marks and setting standards, and chair of the European Network of Forensic Science Institutes' expert working group on marks, the principal scientist at the FSS responsible for Care Assessment and Interpretation and from the UK's Forensic Science Regulator. The likelihood ratio was strongly recommended by the above experts and also by a growing number of forensic scientists within the United Kingdom.

The Forensic Science Regulator has gone on to suggest that it was illogical to say the Bayesian or likelihood ratio approach could be used in some areas of scientific evidence, but not in others. The Appeal Court adopted a different position stating that 'we do not agree with the observations of the Regulator that a similar approach is justified in all areas of expertise'. In response to the ruling in R v T, many forensic scientists have voiced their opinion that the Likelihood-ratio framework remains the most appropriate framework for the evaluation of forensic science evidence, even signing a statement to this effect.

Universal Standards and Best Practice

The National Academy of Science published a report 'Strengthening Forensic Science in the United States: A Path Forward' which outlined the challenges facing the forensic science community, and made recommendations which aim for universal standard and best practice. In 2005, the Science, State, Justice, Commerce and Related Agencies Appropriations Act of 2006 came into force. Under this Act, Congress authorised the National Academies of Sciences to conduct a study on forensic science, stating "While a great deal of analysis exists of the requirements in the discipline of DNA, there exists little to no analysis of the remaining needs of the community outside of the area of DNA" Although in theory the Committee was not given the task of specifically studying DNA evidence, the very essence of it is entwined throughout the report.

The Committee were instructed to focus 'outside of the area of DNA' and instead focused their attention on alternative forensic science techniques such as fingerprint examination. Although DNA may not be at the forefront of the NAS Report, the role which forensic DNA analysis has played in the history of forensic science is evident throughout. The long anticipated report largely states what academics and scientists have noted for some time – that the state of the science used in forensic science and thus in the criminal justice system is extremely poor.

The overriding message of the NAS Report was that forensic science has failed to meet the demands of science. Essentially, the Committee called for an overhaul of the practice and procedures of many forensic science disciplines, including fingerprint analysis, hair and fibre evidence and the identification of shoe prints. The NAS Committee proposed 13 recommendations. One of the main recommendations of the NAS Report was that Congress establish an independent federal entity, the National Institute of Forensic Science (NIFS), to oversee and establish standards, research, forensic practices, education, certification, accreditation and development of technology.

Other recommendations included establishing standard terminology and reporting procedures and to conduct research to address issues of accuracy, reliability and validity. Congress has not yet acted upon the recommendations of the NAS Report, but the findings of the report have echoed throughout the forensic science community and parts of the legal community. Formally, the scientific organisations have supported the general recommendations of the report whilst maintaining the belief that future scientific research will validate most of the bases of forensic science discipline. In the aftermath of the NAS report, it is clear that cases that rest wholly or partly on forensic evidence must be carefully scrutinised.

Complex scientific evidence increasingly plays a role in many of today's court rooms estimony from forensic DNA analysts can overwhelm jurors who often have limited knowledge of the concepts of biology, genetics and technology used to generate DNA profiles. This then raises the question: If jurors do not understand the science underpinning the evidence, how can they then correctly interpret the evidence once it has been presented in court? New Zealand scientists John Buckleton and James Curran have said: "There is a considerable aura to DNA evidence. Because of this aura it is vital that weak evidence is correctly represented as weak or not represented at all."

Where there is uncertainty in the data (e.g. with complex DNA mixtures, partial profiles or low level DNA templates) this should be emphasised so that the jurors may appropriately evaluate the weight that should be given to the evidence which has been presented. It is apparent that there is still a great deal of confusion regarding the presentation of statistical evidence to juries and in particular the field of DNA still requires a great deal of analysis. The inherent danger of this forensic science technique arises if the scientific data underlying the DNA results has been inaccurately presented to the jury. More importantly what influence can these potentially incorrect perceptions regarding scientific data have on a juror's decision making ability?

THE LAY INTERPRETATION OF DNA EVIDENCE

Given the scientific admissibility and reliability of this technique, DNA is frequently used in Court therefore this thesis would not be complete without discussing the lay person interpretation of DNA evidence. Over the past 30 to 40 years, jurors have been criticised for their interpretation of complex scientific evidence. Many critics have questioned whether a jury of inexperienced and untrained people can reach the correct decision in trials which require comprehension of substantial qualities of complex scientific evidence. The presence of forensic science evidence, particularly DNA evidence, predicts convictions. Furthermore, much of the critical attention is focused on so called "first generation techniques" – methods such as handwriting analysis, ballistics and tool or bite mark analysis. Forensic DNA analysis is often excluded from critiques of forensic methods as it is assumed to exemplify the kind of scientific rigor that first generation techniques have not been subjected to.

A lay person who is only slightly acquainted with forensic techniques may suspect that "first generation" techniques are not as scientific or objective as forensic DNA analysis and this could prove dangerous when interpreting DNA evidence. The jury's interpretation of this complex scientific and technical evidence is key in ensuring justice is done. As the admissibility of certain DNA profiling techniques has become less controversial, an important question remains: Are jurors able to understand the complex scientific and inherently probabilistic testimony that accompanies a DNA match report?

SCIENTIFIC EVIDENCE & THE JURY

DNA evidence is a crucial predictor of guilty verdicts in jury trials. Juror difficulties in understanding and applying the scientific and statistical information conveyed by forensic experts about a DNA match have been documented in a number of studies. Post-trial interviews of jurors in six criminal trials revealed that jurors who admitted to difficulty in understanding DNA expert evidence presented convicted despite the lack of understanding. Jurors often have high expectations of scientific evidence with a study held in 2008 stating that 46% of jurors "expected to see some kind of scientific evidence in every criminal case."

The research also found that 22% expected to see DNA evidence in every case. One Court described this troubling aspect of scientific evidence as its ability to assume "a posture of mystic infallibility in the eyes of the jury." The very fact that the judge has admitted DNA evidence into the courtroom often leads to juror assumption that the evidence must be reliable and therefore jurors tend to rely on the results without thoroughly questioning it and this is where the danger lies.

'WHITE COAT SYNDROME'

A juror can appear to be susceptible to 'White Coat Syndrome' if when faced with DNA evidence in a criminal trial, they are inclined to believe 'science does not lie'. This belief generates confidence and leads to DNA evidence being regarded as infallible. According to new reports from The Australian Institute of Criminology Research jurors are often not confident enough in their knowledge of DNA evidence which may often go against the accused when a guilty verdict is returned. Professor Jane-Goodman Delahunty says "They tend to regard it as infallible and so once it comes in, the very fact that the judge has admitted the evidence is often construed as an indication that it must be reliable, and therefore rely on it without questioning it very thoroughly." Without questioning the science underpinning the evidence, it is difficult to correctly interpret the evidence taking into consideration the fragilities of the technique.

Professor Delahuntly conducted a study of around 400 people to find out what their response would be to DNA evidence in mock murder trials. The study found that people with a low understanding of DNA convicted at a rate of 75%, while those with a better knowledge had 42% conviction rate. She noted; "The conviction rate is far higher than the scholars think it ought to be, if you look at the weight that the evidence might warrant."

It has been questioned whether jury innovation trials could improve juror understanding of DNA evidence. The National Institute of Justice conducted a study on the impact of jury trial innovations upon mock jurors. Although this particular study was focused on mitochondrial DNA this would have broader application to nuclear DNA.

The research showed that jurors used three of the innovations the most—the multipurpose notebook, note taking, and the mtDNA checklist. The multipurpose notebook was the most popular innovation: 92% of the jurors said that the notebooks—in particular, the expert's slides—helped them to remember and understand the case. The second most used innovation was juror note taking with 88% of jurors who took notes. Two-thirds said their notes helped them remember the evidence.

The third most used innovation was the mtDNA checklist: 85% of jurors allowed to use the checklist said they reviewed it during deliberations. Most found that the checklist increased their understanding and recall of the evidence. The least used innovation was jury questioning: only 22% of the jurors allowed to ask questions actually did. Based on the study, researchers believe that the use of certain jury innovations has the potential to improve jurors' comprehension of mtDNA and other scientific evidence.

Methods that provided direct guidance or additional expert information—such as the mtDNA checklist and the juror notebook—best improved juror understanding. This suggests that other jury innovations that provide a better understanding of expert evidence such as juror tutorials in complex subjects and court-appointed experts to discuss the parties' often conflicting scientific evidence are ripe for evaluation.

The results of the study showed that most juries are capable of comprehending and using different forms of DNA evidence at trial. Nonetheless, researchers acknowledged that some jurors are likely to have trouble with complex DNA evidence.

THE 'CSI EFFECT'

A more modern interpretation of white coat syndrome is 'the CSI effect' which has been described as "... the perceptions of the near-infallibility of forensic science in response to the TV show". There are a number of differences between the fact and fiction of forensic science evidence and this has been seen to have a variety of effects on American Society. One forensic scientist, Thomas Mauriello, estimates that only 40% of the so called "science" on CSI does not exist and the rest is an unrealistic portrayal of what actually happens in crime labs.

There are a number of theories surrounding the effect of CSI style programmes however; the one that is heard most often is that CSI has raised the expectations of the public as to what type of evidence may be offered at trial to such a level that jurors are disappointed by the real evidence presented. As a result, jurors are more likely to acquit an accused on the basis that there is not enough forensic science provided in court to persuade them of guilt. Most recently, the "Tech-Effect" theory has replaced the "CSI-Effect" theory as a means to explain the potential impact of technology on jurors. This theory argues that crime dramas do not influence jurors; rather, jurors have heightened expectations for technical and scientific evidence simply because technology is so advanced in today's society.

It is difficult for jurors to comprehend the complex scientific and inherently probabilistic testimony accompanying a DNA match report. The perception that DNA is almost infallible and conclusive in all respects has a significant impact on a juror's ability to correctly interpret the results of forensic DNA analysis. Phenomena's such as the 'White Coat Syndrome' and the 'CSI Effect' have been found to have a significant effect on whether a jury will choose to convict or not. Jury innovation trials appear to suggest that improving the clarity of the evidence and allowing aids can improve juror understanding with regards to complex scientific evidence.

It would appear that the natural progression to improve juror understanding should come in the form of jury innovation trials which allow for a better understanding of expert evidence such as juror tutorials in complex subjects and court-appointed experts to discuss the parties' often conflicting scientific evidence. There should exist a greater awareness of forensic DNA analysis as suggested in the Caddy Review when it stated that the results of LCN DNA should be "reported to the jury with caveats" and that the results of LCN must be interpreted cautiously and all factors regarding the extreme sensitivity of the technique to be taken into consideration by the jury.

CONCLUSIONS & RECOMMENDATIONS

DNA analysis is an extremely powerful scientific tool which has proved to be invaluable to both the forensic science and legal communities. Forensic DNA analysis for identification purposes affords several advantages to law enforcement and legal systems which make possible both apprehension of criminals and exoneration of those wrongly convicted. There are many advantages of this technique: it can be said to be a more reliable form of evidence than eyewitness testimony which was previously used and it provides a scientific basis which allows for a physical link to be made between a criminal and a crime scene in order to secure a conviction.

The availability of such a technique has also allowed for the introduction of Innocence Projects which have used forensic DNA analysis to help exonerate those who have been wrongly convicted. The magnitude of Innocence Projects worldwide success is a reminder of the significant benefits which DNA profiling has brought to the criminal justice system, however, the expansion of new techniques has increased the sensitivity of forensic DNA analysis and as a result the dangers of incorrect interpretation are now more significant than ever before.

The process of forensic DNA analysis is very important and if the results are not reliable, accurate, valid and reproducible then how is it then possible to interpret the results correctly? The validity of one such new technique, LCN DNA, has been a controversial matter. There are a number of risks which accompany the LCN technique and it is clear that special consideration must be given when interpreting LCN DNA results. A number of factors such as contamination and stochastic effects make the interpretation of LCN DNA results very difficult.

In the case of Hoey, Weir J emphasised the cardinal principle of criminal law; "Justice 'according to law' demands proper evidence... "evidence which is so convincing in truth and manifestly reliable that it reaches the standard of proof beyond reasonable doubt." The evidence in Hoey, which based primarily on LCN DNA evidence, failed to meet such an immutable standard. If the dangers of LCN DNA are such that it becomes impossible to tell where such a minute sample came from then how can it be deemed a reliable and valid scientific technique?

The judgement in Hoey led to a review of the LCN technique and a review by the Crown Prosecution concluded that LCN DNA analysis should remain as "potentially admissible evidence." It went on to state that at that time there was no reason to believe to that there is any "inherent unreliability" in the LCN process provided that it is carried out according to the prescribed processed and that the results are properly interpreted. However, the nature of LCN DNA samples make reproducibility a significant problem for obtaining consistent profiles – if the results are inconsistent how then is it possible to give a reasoned interpretation of them?

The Caddy Review also concluded that the LCN technique is sound again emphasising that all the appropriate conditions are met. The Caddy Review also empahsised that the LCN DNA profiles should be "reported to the jury with caveats." It is clear that in order to be accurate, the results of LCN must be interpreted cautiously and all factors given the extreme sensitivity must be taken into consideration. Despite both reviews giving LCN a clean bill of health, there still exists a great deal of scientific ambiguity with regards to the general uses of the technique. In the mind of a juror, however, there does not exist a great deal of difference between standard DNA analysis and the more sensitive techniques such as LCN – DNA continues to predict convictions and this is where the problem lies.

The Law Commission Consultation Paper 190, 'The Admissibility of Expert Evidence in Criminal Proceedings in England and Wales: A New Approach to the Determination of Evidentiary Reliability' reiterated that there is a growing dissatisfaction with the law relating to the admission of expert evidence. The Paper suggested that there should be an explicit "gate-keeping" role and that the onus is on judges to protect the jury from unreliable scientific evidence. The Law Commission's proposal on expert opinion evidence has since been criticised in that it places an onerous demand upon judges, who have the difficulty of interpreting the criteria.

If interpreted too strictly, this could lead to the exclusion of forensic testimony becoming widespread. On the other hand, if interpreted too liberally, the criteria will not provide adequate safeguards against admission of unfounded evidence. A major issue with this proposal is that the scientific nature of forensic DNA analysis has the potential to confuse not only the jury but also the trial judge which could lead to the judge admitting scientific evidence which is unfounded.

It is essential for the forensic science community to correctly analyse and interpret the results and to present truthfully to the jury in a language which they understand. There are many difficulties which face the forensic science community when it comes to the interpretation of DNA results. It is essential that the science underlying the expert evidence is sound. In 2009, The National Academy of Science published a report 'Strengthening Forensic Science in the United States: A Path Forward' which outlined the challenges facing the forensic science community. Although this report placed DNA evidence on a pedestal, it identified that the state science in forensic science and the criminal justice system is very poor. If the science underlying the results is not sound then how can it be viewed as reliable?

Complex scientific evidence can be difficult for jurors to comprehend which makes their task of interpretation more difficult. If jurors do not understand the science underpinning the results, how can they correctly interpret the evidence once it has been presented in court? The jurors have the final say in the interpretation of the results of DNA evidence and accurate interpretation of this complex scientific and technical evidence by the jury is key in ensuring justice is done. A number of studies have documented that jurors have difficulty in understanding and applying the scientific and statistical information and jurors have admitted to convicting despite the lack of understanding of the evidence. It has been questioned whether jury innovation trials could improve juror understanding of DNA evidence and results show that innovation trials do have the potential to improve jurors comprehension of complex scientific evidence which would allow jurors to take more consideration before placing too heavy a reliance on the results.

Although the dangers of forensic DNA analysis are routinely recognised, little seems to have been done to improve the reliability of the technique. To ensure justice is done, it is important for correct interpretation at every stage of the process – from the forensic scientists who analyse it, to the judges who make a decision on the admissibility of it, to finally the jurors who make a final decision based on the results. At present, it is clear that there exists too heavy or simplistic reliance on the results. Forensic DNA analysis is a technique which, although widely accepted by the scientific and legal communities, still requires a great deal of analysis.

There are a number of recommendations which could improve the reliability of DNA evidence. Firstly, more research into LCN and its scientific reliability, reproducibility and validity could improve the use of the technique and ensure that its limitations are clear. Secondly, more research into juries could be conducted to see how they respond to current DNA evidence and how presentation of the evidence can be improved to increase comprehension and to reduce the CSI effect. Jury innovation trials have already been seen to improve juror understanding of this complex scientific evidence.

More public education on the topic could also work to improve awareness of the limitations of scientific evidence, especially DNA evidence. Thirdly, more training for lawyers and the judiciary to ensure all scientific evidence is properly scrutinised in court, in line with the Daubert-style gate-keeping role and the recommendations of the Law Commission. The above recommendations, if implemented correctly, have the potential to improve the reliability of DNA evidence and to prevent future miscarriages of justice.

DNA EVIDENCE AND ISSUES

Advances in technology have made DNA testing an established part of sexual assault investigation and prosecution, especially for cases in which identification is the primary issue. Moreover, these advances have rendered traditional serology identity testing for blood, saliva, and semen virtually obsolete.

- In 1996, for example, there were more than 17,000 cases involving forensic DNA in the United States alone. Most of these were sexual assault cases involving DNA testing of specimens collected from vaginal swabs and semen stains.

The purpose of this module is to outline the potential contribution of DNA technology to the investigation and prosecution of sexual assault cases. We will also explore a number of the barriers we must first overcome to realize this potential, before concluding with specific guidelines for the investigator in collecting, analyzing, and using DNA evidence.

NATURE AND PURPOSE OF DNA EVIDENCE

DNA is found in all bodily fluids and tissues. In fact, it is found in every single cell of a person's body, and each cell has identical DNA. Because of this, DNA evidence collected from a crime scene can be used like a fingerprint to include or exclude a suspect in a particular case. It can also be used to link crime scenes either locally or on a state or national level.

The primary purpose of DNA evidence in a sexual assault case is thus to identify an assailant and/or to prove sexual contact. Experienced investigators often say that they rejoice when a suspect denies engaging in sexual contact with the victim, because the case then provides the opportunity to make full use of forensic technology.

As we've stated repeatedly throughout this manual, however, the vast majority of sexual assaults are committed by someone known to the victim. In these cases, identification is not the issue because the defense will probably claim that the sexual activity was consensual. The value of DNA evidence in these cases where the victim and suspect know each other is therefore minimal.

A BRIEF HISTORY OF FORENSIC DNA

The first sexual assault defendant was convicted in this country with the help of DNA evidence in 1987. The case was upheld on appeal the following year. Since that time, DNA analysis has been used in over 20,000 cases resulting in conviction. Historically, DNA technology has been used primarily as a prosecutorial weapon rather than an investigative tool. In other words, DNA evidence has generally been used to confirm the identity of someone already under suspicion, rather than assisting in the investigation and identification process.

- There are, of course, exceptions where DNA technology has identified a suspect that would have otherwise remained undetected. For example, "in 1991, the Minnesota Bureau of Criminal Apprehension (BCA) Laboratory became the first state crime lab to identify a suspect on the basis of DNA alone."
- An article included in the supplemental materials also describes the success of DNA matches in Florida and Virginia.

Several recent, highly publicized cases have also illustrated the potential for DNA to exonerate individuals wrongfully convicted of rape or homicide. For example, DNA technology has been used to free a number of men in prison for sexual assault, including 12 in 1992 and 28 in the next two years. In response to this situation, Attorney General Janet Reno created the National Commission on the Future of DNA Evidence. The official mandate of this

Commission is to submit recommendations that will help ensure more effective use of DNA as a crime fighting tool and foster its use throughout the criminal justice system.

- This includes using DNA evidence in cases where a suspect is known, by confirming the suspect's presence at the crime scene.
- It also includes cases in which the suspect is not known, and DNA evidence can be compared with known samples that have been collected from previous cases and catalogued in a database.

What DNA (mostly) replaced: Trace evidence

Trace evidence includes items such as hairs, fibers, paint chips, glass shards, shoe prints, gun shot residue, arson/explosives and physical matches. Using this type of evidence, forensic scientists have been able to identify the source only on the basis of its general appearance and structural features. A sample report on such trace evidence is found in the supplementary materials.

- Unlike DNA, trace evidence rarely provides definitive identification. As a result, trace evidence is primarily useful only in cases that do not have DNA evidence; otherwise substantial resources can be wasted by crime laboratories screening for trace evidence that will never be analyzed.

This may change, however, with new technologies. For example, the Office of Justice Programs announced in February of 2000 the release of a publication addressing how human hair found at a crime scene may be matched with samples taken from a specific individual. Of course, analysis for trace evidence will continue to be used for identification purposes in isolated situations.

It will certainly be useful in those cases lacking DNA evidence. However, it can also be used in another way. For example, if both DNA and trace evidence are found in a sexual assault case, the DNA will be used to identify the assailant. In some of these cases, other assaults might be seen as related because of a similar modus operandi, suspect description or geographical area. If evidence collected in those cases does not contain relevant biological material, the trace evidence collected in the first case might be used to link it with both the other cases and the definitive identification provided by the DNA.

Finally, trace evidence can always be used for purposes other than identification. Trace evidence can be used to corroborate the victim's account of events, by linking the suspect with particular locations. For example, even if the suspect admits having sex with the victim, he may deny certain aspects of the event. If these aspects can be corroborated with trace evidence, this bolsters the credibility of both the victim and her account of what happened.

Sources of DNA evidence

The promise of DNA evidence for sexual assault investigation lies both in its widespread availability and longevity, as well as its unrivaled potential for identifying a suspect or proving that sexual contact took place. At the crime scene, DNA evidence can be collected from many kinds of biological specimens including blood, saliva, semen, hair, skin cells, bone, teeth, tissue, urine, feces, and vomit.

- For example, a suspect's DNA may be collected from saliva found on the victim's body or clothing; it can even be found on the victim's bra if she puts it on after the suspect has kissed, licked, or sucked on her breasts.
- DNA evidence can also be collected from anything that the suspect put in his mouth, including chewing gum, cigarette butts, envelopes, and drinking cups. It can even be found on articles of clothing, such as hatbands.
- Fingernail scrapings or swabs from a victim's hands can additionally be used to collect DNA evidence transferred from the suspect, and from the suspect's hands to corroborate digital penetration of the victim's genitals.

The potential for collecting DNA evidence from the victim, suspect, and crime scene is almost unlimited. Moreover, the DNA molecule is long-lived and likely to be detectable for many years in bones or bodily fluids. This means that old cases can now be solved and possibly prosecuted using current forensic technology.

BARRIERS TO REALIZING THE POTENTIAL OF DNA EVIDENCE

Despite the exciting promise of DNA technology, a number of barriers remain to realizing its full potential. One of these barriers is the frequent failure of law enforcement to identify and collect appropriate DNA evidence from the crime scene. Many law enforcement agencies have not been properly trained to recognize and collect potential DNA evidence, and this situation leads to an unnecessary disadvantage for the investigation and prosecution of sexual assault cases. For example, a recent FBI survey revealed that of all sexual assault cases, less than 10 percent had DNA evidence submitted to crime laboratories.

Other barriers include the failure to effectively evaluate DNA evidence for analysis, lack of communication between law enforcement and crime lab personnel, limited resources, and the use of incompatible systems for DNA analysis. As summarized in the National Institute of Justice Journal, "despite the abundance of evidence, and despite the advantages of DNA testing, little of this evidence is recovered from crime scenes, less is submitted to crime labs, and still less is analyzed."

Failure to effectively evaluate DNA evidence

When analyzing evidence for DNA, processing a pure sample such as blood or a saliva swab is only a small part of the process. Much of the evidence with DNA potential is not pure but rather collected from the crime scene (from clothing or bedding, etc.). The problem with this type of evidence is that it requires effective evaluation by law enforcement in order to provide information to assist crime lab personnel in their analysis.

- For example, one sheet or shirt collected from a crime scene may contain numerous stains. If the suspect ejaculated on the victim's shirt, this should be noted. Similarly, if the victim states that semen from the suspect can be found at the foot of the bed, the sheet can be oriented to indicate which is the head and foot. All of this information will be useful for crime lab personnel who are screening the evidence.
- Once a stain is detected, moreover, it must be extracted for DNA comparison. This continues to be a time consuming and costly process and any information provided by law enforcement to guide this process is extremely valuable.

Unfortunately, law enforcement has traditionally received very little training in how to evaluate potential evidence in this way. Patrol officers are taught to impound everything they believe might be associated with the crime scene – and this is a good thing given the time sensitive nature of evidence collection. However, this creates a situation in which it is not uncommon to have 10 or more large bags of evidence impounded in the property room (e.g., sheets, blankets, sofa cushions, clothing, sexual assault evidence kit).

To make things worse, detectives are rarely taught about the process used by the crime lab to analyze evidence or how different types of crime lab personnel (e.g., those specializing in trace evidence, latent prints, or biological evidence) might be assigned to a case.

- As previously stated, trace evidence includes items such as hairs, fibers, paint chips, glass shards, shoe prints, gun shot residue, arson/explosives and physical matches.
- In contrast, forensic biology (a separate unit within many crime laboratories) will analyze DNA from blood, semen, saliva, fingernail scrapings, etc.

Given this lack of understanding, it is common for detectives to submit a lab request to "analyze all evidence for trace and semen." Neither detectives nor their supervisors typically understand the process set in motion as a result of this request.

Lack of communication between law enforcement and crime laboratory

Just as police officers often fail to understand how to effectively evaluate evidence for analysis, a traditional lack of communication with crime lab personnel has also limited the contribution of DNA technology. For example, in many communities patrol officers, detectives and prosecutors rarely speak to crime lab personnel. Furthermore, when they do speak they are often unable to understand each other because law enforcement and crime labs speak entirely different languages.

In addition, some have noted that the ease with which DNA evidence can be collected (and in the future, analyzed) will only make it easier to taint or plant such evidence. For this reason, it becomes even more critical for law enforcement and crime lab personnel to appropriately handle and document the chain of custody for all DNA evidence.

Limited resources

In addition to these problems that result primarily from a lack of appropriate training and communication, both law enforcement agencies and crime laboratories suffer from limited resources that further hinder the contribution of DNA technology. This situation is especially pronounced for sexual assault, as these cases typically make up the majority of DNA work performed.

As a result of limited resources, recovered DNA is tested in only 6% of the 250,000 sexual assault cases investigated nationally, leaving a backlog of over two hundred thousand cases awaiting processing.

- This reality often comes as a surprise to officers and others, especially given the recent emphasis on offender databases resulting from Megan's law and other developments.
- In particular, rape crisis advocates, victim rights groups, and medical personnel are often shocked to realize how few sexual assault cases are actually analyzed for forensic evidence.

Those working in the field of sexual assault recognize how difficult it is for a victim to go through the process of a forensic examination. Given the difficulty of the process for victims and the cost of the forensic examination (typically ranging from $200 to $700), it is a tragedy that most evidence collected in sexual assault cases is never analyzed.

Use of incompatible systems for DNA analysis

To further complicate matters, even when evidence is appropriately collected, screened, and analyzed for DNA it can be limited in its contribution by the use of incompatible systems. Forensic laboratories across the country have used different DNA testing systems, including DQA1, Polymarker, RFLP, PCR, and STR. Labs will sometimes even utilize one analytic system for crime scene evidence and another for the suspect's reference standard. Results are therefore frequently found to be incompatible with each other and/or with the state databanks or CODIS.

- To briefly summarize some of the different analyses for DNA evidence, RFLP stands for Restriction Fragment Length

Polymorphism. The advantage of RFLP analysis is that it is informationally rich, but it takes a large quantity of specimen to conduct and about six weeks to process. For these reasons, RFLP testing will likely be supplanted by PCR analysis.

- PCR stands for Polymerase Chain Reaction, an analysis which takes only days to perform and can be conducted with a smaller quantity or poorer quality of specimen. PCR produces much less information, however, in comparison with traditional RFLP analysis.
- More recently, STR (Short Tandem Repeats) technology has been developed which can produce as much as information as RFLP tests. More importantly, current STR marker sets can be conducted with small and/or degraded DNA specimens, and the analysis also takes only days to complete.

Although technology is improving, processing DNA evidence remains a time consuming and costly part of sexual assault investigation. As a result, it is imperative that patrol officers, detectives and prosecutors understand what evidence should be collected and examined, based on the history of the assault and how DNA results may affect the outcome of the investigation.

OVERCOMING THE BARRIERS

Thanks to the development of new methods of analysis, the ability of crime labs to process DNA evidence within a reasonable time is expected to improve substantially within the next few years. Regardless, forensic laboratories must have sufficient resources to analyze the evidence submitted and law enforcement must learn to use these precious resources wisely. To accomplish this task, it is recommended that agencies review their policies regarding lab service requests, and make any necessary refinements in direct consultation with crime lab personnel.

- For example, lab service request forms can be revised to include preliminary information about the case. Using the specific details and history of the sexual assault case being investigated, the form can be revised to help detectives focus on what service is being requested from the crime laboratory and why it is being requested. A sample form for this purpose is included in the supplementary materials, entitled "Preliminary Rape Case Information."
- This kind of form can also be used to provide crime lab personnel with guidance to assist in their screening and analysis -- based on the history of the victim and the nature of the assault. Ultimately, an improved lab service request form can assist both law enforcement and crime lab personnel with determining which item(s) will most likely contain potential DNA evidence. A sample of such a service request from crime lab personnel is also provided in the supplementary materials.

Another development that will hopefully assist in overcoming the barriers to DNA potential is the move toward uniformity in analytic techniques. In 1994, the Violent Crime Control and Law Enforcement Act was implemented. Among other things, this Act advocated that uniform standards be used for forensic DNA testing. It further provided federal support for state and local law enforcement agencies to improve their DNA testing capabilities – with the ultimate goal of facilitating their participation in CODIS.

- To establish compatibility among laboratories, the FBI has also promulgated a core set of RFLP and STR loci (specific places in DNA) to be used in forensic analysis.

Of course, databases such as these will only be useful when evidence collected from crime scenes is appropriately identified, collected, screened, and analyzed. Specific guidelines to assist in this process follow.

SPECIFIC GUIDELINES FOR THE SEXUAL ASSAULT INVESTIGATOR

A number of suggestions have already been offered in this module to assist the sexual assault investigator in making full use of forensic DNA technology. For example, by appropriately collecting and evaluating evidence the investigator can provide much needed guidance to the crime lab personnel for DNA analysis. By improving communication and lab service request forms, law enforcement can more effectively assist the crime laboratory by summarizing details from the specific case to determine which item(s) are most likely to have evidentiary value.

- Obviously, all of these improvements depend on collecting extensive information about the assault history from the victim, as described in the module on victim interviews.
- It will also depend on effective communication with medical personnel who conduct the medical interview and forensic examination (see the module on the forensic exam).\

Blood evidence

If blood evidence is collected, the crime lab will need to know whether the victim and/or suspect bled during the assault. If so, it is also necessary to determine which areas of the body produced blood and whether the victim was menstruating at the time of the assault. Finally, the crime lab will need to know whether the victim or suspect received a recent blood transfusion (i.e., in the last 90 days) that might interfere with the DNA results.

CLOTHING EVIDENCE

The investigator will also need to determine whether clothing was collected from the scene and whether it was worn during or after the assault. In most cases, all or part of the victim's clothing is removed during a sexual assault.

This clothing should therefore be identified as it could contain preejaculate fluids, saliva stains, or trace evidence:

- After the assault, clothing is often used by the suspect and/or victim to wipe the genitals following a sex act. It is critical to identify and collect this clothing, along with any that is put on after the assault – especially those pieces of clothing worn closest to the genital structures.
- Most officers are not taught to identify this clothing separately, but this information is crucial for the crime lab when screening a large amount of evidence.

Basically, all of this information collected from the victim and passed along to the crime laboratory can help to determine which item(s) of clothing will most likely contain seminal fluids, saliva, or other evidence with DNA potential.

Bedding evidence

If bedding is collected, similar questions must be asked of the victim to determine which item(s) will most likely have seminal fluid or other evidence with DNA potential. Throughout this process it is important to remember that the purpose of these questions is to reduce the demands on crime lab personnel to screen large amounts of evidence.

- For example, if the assault occurred in the suspect's bed, the presence of his own seminal fluid will be irrelevant to the case under investigation. However, the presence of the victim's epithelial cells would at least place her in the suspect's bed in the case of a denial.
- If the assault took place in the victim's bed, every effort must be made to identify stains as relevant versus irrelevant to the assault. For example, if the victim previously had consensual intercourse in the bed, these stains must be identified as resulting from the prior partner(s).

If the victim and suspect engaged in consensual sex in the same bed prior to the assault, any analysis for semen or the victim's epithelial cells would be a waste of valuable crime laboratory resources.

Collecting reference standards

One often overlooked reality about DNA evidence is that is only meaningful when considered in the context of reference standards.

- For example, DNA evidence collected from the victim's body can only be identified as the suspect's if a reference sample has been obtained from the suspect.
- Similarly, semen recovered on the victim's body or bedsheets can only be identified if reference standards are collected -- not only from the suspect but also from any previous consenting partners whose semen could still be present.

As a result, detectives need to ask the sexual assault victim if she had consensual sex within 96 hours (4 days) prior to the time the biological samples are collected. When clothing or bedding are collected, the detective must also determine if seminal fluid stains might be present from previous consensual partners. In either case, reference samples must be collected from all consensual partners whose semen or other DNA evidence might still be present. This should be done before submitting a DNA request to the crime laboratory.

TYPES OF REFERENCE STANDARDS

Blood and saliva are the most common source of reference standards for DNA testing. Thus, samples from the suspect can be collected either through a forensic examination by medical personnel or with mouth (buccal) swabs that can be obtained in the field by law enforcement.

- Mouth swabs are becoming increasingly common in sexual assault investigation, both because they are easy to obtain and because many suspects will readily consent to cooperation.
- Mouth swabs are also non-intrusive, as they do not require transporting the suspect to a medical facility or sticking a needle in his arm.

However, the increasing popularity of this technique requires that both patrol officers and detectives be trained to carry the appropriate equipment and properly obtain samples for DNA testing.

SUBMITTING THE LAB SERVICE REQUEST

Once the preliminary rape case information has been completed and reference samples collected, the detective is ready to submit the lab service request. When doing so, successful sexual assault investigators must prioritize those items of evidence that should be analyzed for trace or biology depending on the type of investigation and the history of the assault obtained from the victim.

- The item listed should be specific to provide the most effective guidance to crime laboratory.
- For example, detectives should request that the "victim's pink underpants" or "vaginal swabs" be analyzed, rather than the "victim's clothing" or "rape kit."

Obtaining supervisory approval

In addition to improving communication with crime lab personnel and revising the forms used for requesting lab services, it is also recommended that supervisors review the lab service requests completed by sexual assault investigators. This secondary review allows supervisors to make sure that only legitimate requests are being submitted, in addition to tracking the number of requests submitted to the crime laboratory. Once this review process is in place, the supervisor(s) can then meet regularly with sex crimes investigators and the DNA Section manager, in order to discuss and prioritize current work in progress, new requests, and any upcoming court dates.

Collecting fetal tissue evidence

Some sexual assaults result in pregnancy, especially those involving adolescent victims. In these cases it may be necessary to collect DNA evidence from the infant if it is carried to full term, or from the fetus if it is not.

- If the victim elects to carry the fetus to full term, a mouth swab or blood sample can easily be obtained from the infant once it is born. Mouth swabs are preferred for this purpose as they are less traumatic for the infant and require no specialized medical assistance.
- However, if the victim elects to abort the fetus, the investigator must immediately work to facilitate collection of valuable DNA evidence.

Depending on the facility chosen to perform the abortion and the number of weeks pregnant, investigators will encounter a variety of responses from medical facilities when attempting to collect fetal tissue samples. Jurisdictions will also vary in whether they require an investigator to have consent or a search warrant before fetal tissue can be collected.

- It is critical that investigators alert the medical facility to not add any fixative or preservative such as formalin, formaldehyde, or any other liquid.

Collection, packaging, and storage of DNA evidence

Obviously, DNA evidence can only be useful to sexual assault investigators if it has been appropriately collected, packaged, and stored. DNA is usually collected from crime scenes in the form of fresh blood or dried blood, fresh liquid semen or dried stains.

- Depending on an agency's resources, these specimens are identified and collected either by patrol officers or crime lab personnel. In some cases, an investigator might be required to conduct evidence collection at the scene of a crime.
- Wherever possible it is encouraged that crime scene processing be conducted by forensic specialists, as they have specialized training and experience as well as access to equipment such as an alternate light source that can be used to detect potential DNA evidence.

Specific guidelines for collecting, packaging, and storing DNA evidence are provided in the Appendix A. Officers are reminded, however, that the entire item containing the stain should be collected wherever possible (e.g., bedding, clothing, mattress, sofa cushion or car seat) to facilitate processing by the crime laboratory.

Conclusions and Future Directions

To conclude, a number of barriers exist that limit the contribution of DNA technology to sexual assault investigations. Given limited resources it thus makes sense for investigators to work smart, using existing resources most efficiently. This means that DNA analysis should currently be conducted only when necessary to establish an element of the offense – primarily for identification but perhaps also to prove sexual contact when it is denied by the suspect.

- In the future, however -- as technology improves and the costs associated with DNA analysis decrease -- it will be advisable to conduct DNA analysis even in cases where consent is the primary issue. In these cases, DNA evidence may not be particularly useful in challenging the consent defense but it could prove invaluable in linking the suspect to prior (or future) unsolved cases.
- Misdemeanor cases (e.g., peeping, indecent exposure) should also be analyzed for the same reason. In fact, there is at least one case of a sexual assault perpetrator being identified solely on the basis of DNA evidence collected at another crime scene in which a misdemeanor case was being investigated.

The importance of this need is best illustrated by the research of Ann Burgess and colleagues, in which 41 serial rapists were interviewed. Taken in combination, these men admitted to having committed 837 rapes and 401 attempted rapes. The critical finding, however, was that each admitted that their earliest victims were younger siblings, neighborhood children, girlfriends, spouses, and other acquaintances.

- This clearly indicates the need to analyze potential DNA crime scene evidence in all cases since a perpetrator's behavior will often cross over from acquaintances to strangers and escalate.

Of course, any advances in the use of DNA technology will require improvements in the status of databanking, as described in Appendix D and supplemental materials. However, it will also include improved training of law enforcement and increased communication with crime lab personnel. It is hoped that this manual can provide one important step toward strengthening this link and fulfilling "the unrealized potential of DNA testing."

GUIDELINES FOR COLLECTING, PACKAGING, AND STORING DNA EVIDENCE

Collecting and Packaging DNA Evidence

1. Use clean latex gloves for collecting each item of evidence. It is recommended to change gloves between the handling of different items of evidence.
2. Each item of evidence must be packaged separately.
3. Blood stains, semen stains, and other types of stains must be thoroughly air-dried and packaged in sealed paper envelopes or paper bags. To dry stains, a hair dryer can be used on the coolest setting. For large amounts of liquid, a fan can be used.

4. Increasingly, condoms are part of the sexual assault crime scene. Used condoms should be collected and placed in a sterile tube. The tube should then be frozen until analyzed. If a sterile tube is not readily available, officers should make sure – at a minimum – that the condom is allowed to air dry before packaging. Several layers of paper bags can then be used for packaging.
5. For proper chain of custody, all packages must be marked with the case number, item number, and date. Packages must also be initialed across the seals.
6. If stains must be transferred from an unmovable surface (such as a window or sidewalk), sterile cotton swabs and distilled water may be used.
   A. Photograph the surface with a ruler before swabbing. Lightly moisten the swab with distilled water.
   B. Rub the stained area with the moist swab until all of the stain is transferred to the swab. If one swab is insufficient to collect all the stain, use additional moist swabs to collect all of the stain.
   C. Two additional swabs should be collected as substrate controls for DNA tests. Swab #1 should be moistened and used on an unstained area adjacent to the stained area. Swab #2 should be provided with nothing else on it but the water used in the collection process.
   D. Prepare properly marked envelopes or paper containers for the swabs.
   E. Air dry the swabs without permitting them to touch one another. If time requires, the swabs may be placed still moist in paper envelopes. (Glass or plastic containers should never be used. Paper containers allow moisture to escape which helps to prevent bacterial degradation of the DNA.)
   F. Place swabs in appropriate separate paper containers, properly marked for identification.
   G. Scraping dried stains instead of swabbing should only be done if the surface is perfectly smooth and the scraping will result in almost no loss of material. For example, a stain on a smooth vertical surface can be collected (after photographing with a ruler in the picture) by folding a clean sheet of paper in half and taping the top edge of the paper to the surface directly beneath the stain. With a sterile scalpel blade or unused single-edged razor blade, the stain can be scraped into the fold in the paper. Then carefully remove the paper from the surface, remove the tape, fold the paper into a packet, seal with evidence tape and initial properly.
7. Evidence which is incapable of drying such as pieces of tissue, organ, bone, liquid urine, vomit, or other biological material should be

packaged separately in an air tight container. The container should be sealed and properly marked for identification, then immediately frozen and kept stored until analysis. Formalin or formaldehyde should never be used to preserve any biological evidence because these chemicals degrade DNA.

8. Known standard blood samples from deceased individuals should be transferred by syringe into a purple top tube. They should then be packaged, stored, and shipped using the same procedures as known standard blood samples from a living person, as described below.
9. Known standard blood samples from living persons should be drawn in purple top tubes, properly marked for identification, placed in a paper container, sealed with evidence tape for proper chain of custody, and stored refrigerated until analysis. If needed, additional packaging should be used for protection from breakage during shipment.
10. Known standard blood samples (either liquid or dried stains) from persons diagnosed with HIV or hepatitis should be shipped in special containers clearly marked on the exterior that they contain HIV or Hepatitis infected blood.

Storing DNA Evidence

1. Most biological evidence is preserved best when stored dry and cold because dryness and lowered temperatures reduce the rate of bacterial degradation of DNA evidence. With the exception of liquid whole blood samples, the colder the storage the better.
2. For dried stains, material should be frozen (-20 degrees C) or refrigerated (4 degrees C) in separate paper containers. Dried stains which are very old and have been stored at room temperature for months or years will obviously not be hurt significantly by additional short-term storage at room temperature. Nevertheless, it is recommended that these samples be stored cold until they can be analyzed.
3. For undried tissue such as bone, liquid urine, or other undryable biological material (except blood standards), these samples should be kept frozen (-20 degrees C) in separate air tight containers. Glass containers should be avoided as they can break when frozen. Formalin or formaldehyde should never be used for storage as they degrade DNA.
4. For liquid blood standards, these samples should be kept refrigerated (4 degrees C) in their original glass tubes. They should not be frozen.

SHIPPING DNA EVIDENCE

1. All samples should be hand carried on a business day or shipped by priority overnight courier service on a business day, but never shipped over a weekend or holiday.

2. All shipments should be accompanied by a case submission letter attached to the outside of the package, addressed to the attention of the forensic laboratory.
3. Dried stains should be packaged separately and shipped in a sealed box or sealed envelope at room temperature by priority overnight courier service. It is preferred to ship only the stained portion of clothing rather than the entire garment. The suspected stain should be clearly marked.
4. Undried tissue: To avoid putrefaction during shipping, tissue, organ, bone, and liquid urine samples (in plastic tubes) should be shipped by priority overnight courier service in a sealed plastic container placed on abundant dry ice in styrofoam containers.
5. All liquid known whole blood standards should be packaged separately in styrofoam tube packages or wrapped individually in bubble wrap and secured with tape. The individually wrapped tubes of blood should be placed in a cardboard box surrounded by styrofoam chips for protection from breakage. Liquid known whole blood samples should be shipped at room temperature by priority overnight courier. Liquid blood should never be shipped on dry ice because the glass tubes will break.
6. Blood standards from persons diagnosed with HIV or hepatitis should be shipped using a Class 6.2/95 CAN/8-2SAF-T-PAK container with the proper Federal Express shipping tag for dangerous goods. Blood standards should be shipped at room temperature by priority overnight courier service.

5

Forensic Tool Requirements

BASIC CUSTOMER REQUIREMENTS

Investigate the computer forensic techniques for improving the accuracy of the evidences on web accesses by correlating the events in the other logs.

Due to the increasing cyber threats and potential insider attacks, it is critical to validate the web accesses are indeed generated by the person of interest and not planted by others.

PURPOSE

The web accesses are typically captured by the cookie files on the client side and the access logs on the server side. The web accesses of a person can be verified by correlating the login period in system access logs. It can also be supported by the access logs of other applications such as, email, instant messaging or Skype.

Of interest in computer crime evidence collections are:

- The date and time of the access to a certain site.
- The frequency of the access.
- Other supporting evidences that the person is using the same machine.
- The collaborating evidences that the web site has corresponding access records.

Investigate how to weigh the different support evidences and related practices by District Attorney's Office Investigators Office.

PROJECT SCOPE

Scope Definition

Develop as.NET windows form application that can be used by the investigators or the forensic technician. The application will allow for the requesting and reporting of the case related forensic information.

Overall System Scope

SQLlite will be used to store the case information. The web interface will be developed using Microsoft Development Suite (Visual Studio C#).

OVERALL DESCRIPTION

Product Perspective

This application will assist the investigator with information about the user's browser activities and history, any other activities using outlook, Skype and instant messaging. The information collected can be used to find out the frequency of a user visiting a particular web site, the history of the user visiting that web site, and other login activities that could be used as evidence against the user.

A zip file would be given to investigators; they can open the file on the computer that an alleged crime was committed, open a case, run the tool and view/save the reports

Product Features

- Report tool: Investigators will have the ability to run reports as needed it. These reports will include a timeline for each activity on the browsers, IM, Skype or Outlook.
- Printing: Case can be saved and then printed
- Database: All data input can be saved into a database
- Entry: New entries can be added to the database through the New case->save
- Reload: Reports can be run on previously saved cases.

OPERATING ENVIRONMENT

The system will be using Windows XP, SQLite, NET 3.5 and a C# compiler.

METHOD

GOAL AND OBJECTIVES

Make a Forensic tool application that looks for:

- Users Profile
- Surfing history
- Typed URLS
- Number of visits to particular sites
- Cookies
- Cache
- Instant Messenger Activities
- Outlook Activities
- Skype Activities
- Timeline for each activity (Report)

DESIGN OF THE FORENSIC TOOL

The Players

- Internet explorer,
- Mozilla Firefox
- Goggle Chrome
- Skype
- Outlook
- MSN Messenger:

Utilities

- SQLite
- Outlook Redemption
- Windows Log Parser
- ChromeCacheView
- IECacheView
- IEHistoryView
- MozillaCacheView

UNDERSTANDING THE REGISTRY KEY (BRIEF)

Registry Hives: (Only two are real, the other ones are shortcuts or aliases to branches within one of the two hives.

- HKEY_CLASSES_ROOT .... HKCR
- HKEY_CURRENT_USER ....HKCU
- HKEY_LOCAL_MACHINE ....HKLM...real
- HKEY_USERS ....HKU......real
- HKEY_CURRENT_CONFIG ......HKCC

Computer accounts, user accounts, groups, and other security-related objects are security principles. Security Identifiers (SIDs) uniquely identify security principles. Each time Windows XP or Active Directory creates a security principle, they generate a SID for it. Windows XP's Local Security Authority (LSA) generates SIDs for local security principles and then stores them in the local security database.

The Domain Security Authority generates SIDs for domain security principles and then stores them in Active Directory. SIDs are unique within their scope. Every local security principle's SID is unique on the computer. And every domain security principle's SID is unique within any domain in the enterprise. What's more, Windows XP and Active Directory never reuse a SID, even if they delete the security principle to which that SID belonged. Thus, if you delete an account and then add it back, the account gets a new SID.

HKEY_USER: HKU contains per-user (user specific) information. HKU contains at least 3 sub keys: Default, SID and SID_CLASSES (contains per user class registration and file association)

- HKU has other well known SID in Windows XP (Wong, 2006)
- S-1-5-18 refers to LocalSystemAccount
- S-1-5-19 refers to LocalServiceaccount
- S-1-5-20 refers to NetworkServiceAccount

HKEY_LOCAL_MACHINE: Contains per computer settings which apply to all users logging into that particular computer

Taking Ownership of Keys

By default, Windows XP assigns ownership to the HKLM and HKCU as follows:

- Administrators own each subkey in HKLM.
- Users own each subkey in their profile hives, HKCU.

If you have full control of a key (and administrators usually do), you can take ownership of it if you're not already the owner (NEED SOURCE):

1. In Regedit, click the key for which you want to take ownership.
2. On the Edit menu, click Permissions; then click Advanced.
3. On the Owner tab, click the new owner

UNDERSTANDING THE ACTORS

User Profile

A user profile describes the desktop computing configuration for a specific user, including the user's environment and preference settings. A profile is created the first time that a user logs on to a computer running Windows Server 2003, Windows XP, Windows 2000, or Windows NT Workstation. A user profile is a group of settings and files that defines the environment that the system loads when a user logs on. It includes all the user-specific configuration settings, such as program items, screen colors, network connections, printer connections, mouse settings, and window size and position. Profiles are not user policies and the user has a profile even if you don't use Group Policy.

Depending on how you manage your network, you or a user can define the desktop settings. The following user profiles are available in Windows Server 2003, Windows XP Professional, and Windows 2000 Professional (TechNet, 2010):

- *Local User Profile*: Created the first time that a user logs on to a computer, the local user profile is stored on a computer's local hard disk. Any changes made to the local user profile are specific to the computer on which the changes are made.
- *Roaming User Profile*: A copy of the local profile is copied to, and stored on a server share. This profile is downloaded every time that a user logs on to any computer on the network, and any changes made to a roaming user profile are synchronized with the server copy upon logoff.

- *Mandatory User Profile*: A type of profile that administrators can use to specify particular settings for users. Only system administrators can make changes to mandatory user profiles. Changes made by the user to desktop settings are lost when the user logs off.
- *Temporary User Profile*: A temporary profile is issued any time that an error condition prevents the users profile from being loaded. Temporary profiles are deleted at the end of each session - changes made by the user to their desktop settings and files are lost when the user logs off.

A primary goal of user profiles is to separate each users settings and data from that of other users and the local computer (Technet2, 2010).

Registry *hive*: The registry is a database used to store computer- and user-specific settings. Portions of the registry can be saved as files, called hives. These hives can then be reloaded for use as necessary. User profiles take advantage of the hive feature to provide roaming profile functionality.

The user profile registry hive is the NTuser.dat in file form, and is mapped to the HKEY_CURRENT_USER portion of the registry when the user logs on. The NTuser.dat hive maintains the user's environment preferences when the user is logged on. It stores those settings that maintain network connections, Control Panel configurations unique to the user such as the desktop color and mouse, and application-specific settings. The majority of the settings stored in the registry are opaque to user profiles settings are owned and maintained by individual applications and operating system components (Technet2, 2010).

A set of profile folders stored in the file system. User profile files are stored in the file system in the Documents and Settings directory, in a per user folder. The user profile folder is a container for applications and other operating system components to populate with subfolders and per-user data, such as shortcut links, desktop icons, startup applications, documents, configuration files and so forth.

Windows Explorer uses the user profile folders extensively for special folders such as the user's desktop, start menu and my documents folder (Technet2, 2010). On Windows Server 2003, Windows XP or Windows 2000, profiles are stored in the C:\Documents and Settings folder

Local Profile - Existing User

- The user logs on. Windows checks the list of user profiles located in HKEY_LOCAL_MACHINE\SOFTWARE\Microsoft\Windows NT\CurrentVersion\ProfileList to get the path to the user's profile.
- The users registry hive NTUSER.DAT is mapped to the HKEY_CURRENT_USER portion of the registry.
- The users %userprofile% environment variable is updated with the value of the local profile folder.

- When the user logs off, the profile is saved to the local hard disk of the computer

INTERNET EXPLORER

During Forensic analysis it is often relevant to parse the information in IE cookies files into a human readable format. Cookies aid forensic analyst during the investigation by providing insight to a suspect's internet activity

Internet Explorer stores its data in one key and has three subkeys within it that holds the majority of useful information:

- HKCU\software\Microsoft\InternetExplorer\Main:stores the user's setting in Internet Explorer. It contains information such as search bars, start page, form settings, etc.
- HKCU\software\Microsof\InternetExplorer\TypedURLs: stores all URLs that a user has typed into the address field of the web browser.
- HKCU\software\Microsoft\InternetExplorer\DownloadDirectory: displays the last directory used to store a downloadable file from Internet Explorer.

IE stores data in: the drive\Documents and Settings\user\profile folders

Folders = Favorites, cookies, history, and Temporary Internet Files

Registry stores Typed URL's, Passwords and protected Storage information

IE History

History tracks the websites visited by the user and includes date/time info on C:\Documents and Setting\<username>\Local Settings\History. History folder contains a master index.dat file that tracks the History The History folder displays icons that represent the weekly/daily history activity. Each of these folders contain an index.dat file

IE Temporary Internet Files:

- Located at drive C:\Documents and Settings\user\Local Settings\.
- Contains an index.dat file that records thwe URL, Filename, Username and Content info
- Provides information about browser activity even if the user deletes their Temporary Internet Files
- Review the Temporary Internet Files for cached Internet emails

Registry – Typed URL's

Most URLs that you visit are saved in the History folder. However, Internet Explorer also saves the last 25 URLs that you typed in the following Registry key: HKEY_CURRENT_USER\Software\Microsoft\Internet Explorer\TypedURLs. The lowest numbered entry is the most recent site visited (url1, url2, url3, etc)

Index.dat file

The index.dat file contains a header that harbors important information about the file's properties. Specifically the header will contain the index.dat file length, the HASH table offset and the internet cache directory names (Jones, 2003). The first field we notice is the file size. The file size is given in the file header immediately following the NULL (0x00) terminated version string.

Immediately following the file size is the location of the HASH table. The HASH table is an array of data that contains entries pointing to the relevant activity data within the index.dat file.

- Bytes 0x20 – 0x23: Location of hash table.
  Hash table is used to store the actual entries
  Beginning of hash table:

After the HASH table offset is a listing of directories that this index.dat file uses to store the locally cached files on the user's computer (Jones, 2003). These directories contain the files that were actually downloaded from the web.

Size: 0x00394000 3751936
Hash Table: 0x00005000
Directories: (null-terminated, 0x50)

MOZILLA FIREFOX

Firefox stores a user's personal information such as bookmarks, extensions, and user preferences in a unique profile stored in files within a special folder on your pc. The first time you start Firefox, it will automatically create a default profile; additional profiles can be created using the profile manager. Profile folders are placed in a common location by default but are named randomly for additional security(e.g. "xxxxxx.default" is the profile folder name for the "default" profile, where xxxxxx represents a random strung of characters).

Starting in Firefox 3 a new file format is used to record browser history information. Rather than storing this information in a flat file using the mork file format, the information is kept in a SQLlite database.

It can run from a portable device... leaving few traces on the computer; It can be run from a CD, leaving behind even fewer traces on the computer, none on the CD A USB can be inserted to write block, also eliminating evidence from being created on the USB when running Firefox from the USB device Firefox stores most of its data in files instead of the registry Easy to find, in individual folders and it is very easy for a user to wipe the folders securely. It stores personal information such as bookmarks, extensions and user preferences in a unique profile, called Profile

Files in Profile.ini

Cache, Chrome, extensions bookmarks.html, bookmarks.bak, mimeTypes.rd, cert8.db, compatibility.ini, key3.db, search.rdf, XUL.mfl, prefs.js, signons.tex, components.ini, cookies.txt, defaults.ini, formhistory.dat, compreq.dat, localstore.rdf, xpti.dat, history.dat, secmod.db Profiles.ini keeps track of the profile file.

Whenever you use the Profile Manager to create, rename or delete a profile in these applications, the changes are reflected in the profiles.ini file. The profiles.ini file is a plain text file, it can be easily openened, viewed and edited if necessary. Profile can be store anywhere (Mozillazine, Profiles.ini, 2007)

Firefox History Files

- File location:
  C:\Documents and Settings\<username>\Application Data\Mozilla\Firefox\Profiles\zb0sttcz.default\places.sqlite.

The following registry path is where Mozilla Firefox info is contained: (Musings, 2007)

HKEY_LOCAL_MACHINE\SOFTWARE\Mozilla\Mozilla Firefox

GOOGLE CHROME

- File location:
  C:\Documents and Settings\<username>\Local Settings\Application Data\Google\Chrome\User Data\Default\Local Storage.
  Google Chrome stores the browser history in a SQLite database, not unlike Firefox. Yet the structure of the database file is quite different.
  Chrome stores its files in the following location:
  C:\Documents and Settings [USERNAME] Local Settings\Application Data\Google\Chrome

The database file that contains the browsing history is stored under the Default folder as "History" and can be examined using any SQLlite browser there is (such as sqlite3).

The available tables are:

- Downloads
- Presentation
- Urls
- Keyword_search_terms
- Segment_usage
- Visits
- Meta
- Segments

The most relevant tables for browsing history are the "urls" table that contains all the visited URLs, the "visits" table that contains among other information about the type of visit and the timestamps and finally the "downloads" table that contains a list of downloaded files (Brainfold, 2010)

Skype

Skype is communications software that allows users to communicate with each other in real time using VoIP, video chat or text chat. It is unique among other IM applications in that Skype runs over a decentralized peer to peer (P2P) network rather than routing all communications packets through a central server or cluster of servers (ISO Consensus Paper: Skype, 2009)

For windows systems, Skype's functionality can be managed at a number of levels. Skype configuration and policy settings are maintained in the following hierarchy:

- HKEY_LOCAL_MACHINE Registry Keys
- HKEY_LOCAL_USER Registry Keys
- XML config. Files in C:\Documents and Settings\ <username>\ Application Data\Skype\
- Skype client user preferences and default

Skype software uses a number of files to store data. These files relate mainly to historical information, call histories, file transfers, messaging sessions, etc. They also cache user profiles. The interpretation of these log files can yield a significant amount of information about communications that have taken place through the software. This note considers the range and format of user data stored on the platform.

Information available in log files

This section details the information available for extraction from Skype logs. Note that the sequence number allows the order of events to be determined, without relying on the resolution of the timestamp. The timestamps give date and time to a resolution of one second.

Calls (e.g. ca11256.dbb)

- Sequence Number
- Time stamp
- Username (remote end)
- Screen name (remote end)
- Duration of call (seconds)
- PSTN number (when using Skypeln or SkypeOut)
- PSTN status (when using Skypeln or SkypeOut)

File transfers (e.g. transfer512.dbb)

- Sequence Number
- User name (remote end)
- Display name (remote end)
- Full saved file path
- Filename
- File size
- Time stamp

Messages (e.g. msg256.dbb or chatmsg256.dbb)
- Sequence Number
- Message content
- Chat ID (groups messages within a chat session)
- Timestamp
- User name (sender)
- Display name (sender)

User profiles (e.g. user1024.dbb or profile1024.dbb)
- Username
- Display name
- Language (2-character ISO code)
- Province / city
- Country code (2 character ISO code)
- Phone number
- Office number
- Mobile number
- Thumbnail image

Voicemails (e.g. voicemail256.dbb)
- Sequence Number
- Username
- Display name
- Time stamp

File-naming convention

Files are stored with a.dbb extension with the filename consisting of a string describing the contents followed by a number which indicates the record length. The minimum record length observed is 256 bytes, with files seen up to 16384 bytes. Items are stored in the smallest length format possible with blank padding to fill any space remaining in the record. Therefore it is quite common to have multiple files with the same prefix and different record lengths.

Skype Log File Analysis Skype Log File Analysis

- Call*.dbb Call history
- Chatmsg*.dbb Chat history
- Profile*.dbb Details of user profiles
- Transfer*.dbb Details of file transfers
- Chat*.dbb Chat history
- Contactgroup*.dbb Unknown
- User*.dbb Local user's profile
- Voicemail*.dbb Details of voicemail messages (no contents)

File format

The Skype log files are stored in a flat-file database format:
- Although what's stored in each record differs between files, the underlying structure is the same (e.g. records, record headers etc).

- Only items of the type indicated by the filename are stored in a file.
- Numeric items are stored in little-endian format. For example 1453 is 0x05AD and would appear as two bytes, 0xAD, 0x05.
- A record begins with a 4-byte record header, which is always 0x6C, 0x33, 0x33, 0x6C (or "1331" in ASCII).
- The record header is followed by a 4-byte, little-endian, unsigned integer which indicates the length of the following data. The filename indicates the maximum number of data bytes to be stored in a record in the file. The record length in the file will be less than or equal to this number. The total space allocated to each record is the maximum data length (given by the filename) plus the header size (8 bytes).
- Blank entries have been observed in files. These include the 4 byte header, but have a data length indicator of 0 with all following data bytes set to 0x00.
- There is no indicator for the end of a record - any space remaining in the record after the data is set to 0x00. The last record in a file is often not padded out.
- Any communications item (call, message, file transfer etc) has a sequence number associated with it. This is used by the software to reassemble the order in which events occurred. This is particularly useful as items of different sizes can often be spread across multiple files.
- It is unknown how multi-byte character sets are handled as standard ASCII is all that has been observed for storing textual information.
- Within a record, and between different pieces of information, delimiters indicate the next data item (Skype Log File Analysis, 2007).

OUTLOOK

Email is one of the most common ways people communicate. From internal meeting requests, distribution of documents and general conversation one would be hard pressed to find an organization of any size that does not rely on email. Studies have shown that more email is generated every day than phone conversations and paper documents combined. Forensic Analysis of email clients and servers has been in the spotlight of civil and criminal cases worldwide and no examination of Document Discovery is complete without requesting, searching and organizing email.

What is a PST file? A PST file (personal storage file) is Microsoft Outlooks file format for email storage. This single file, is a compound file, like a ZIP file, and can contain thousands of emails, contacts, tasks, and calendar entries. To view the contents of a PST file, the file can be mounted in Outlook or specialist forensic tools such as EnCase or unmanm,aged code. PST files are important, if not critical, in forensic and electronic discovery investigations as they provide one of the primary storage methods for email within companies.

INSTANT MESSENGER

MSN Messenger has the ability to keep an indefinite log of all conversations. In MSN Messenger select the Tools menu, Options... menu item, and then the Messages tab. At the bottom of the dialog is an item labeled: Message History. Select it and your conversations will be recorded in the directory listed in the box below.

When conversations are saved in the listed directory you'll find several files, typically one per person that you've had a conversation with. The files will all end with ".XML". Not to worry, Internet Explorer can open these files and will display them properly Is there any other way to retrieve history if the "auto archive messages" options are not chosen?

Not that I'm aware of, anyway. Conversations, when not being archived, may never even hit the disk, so there's nothing saved to be recovered. There might be small, tiny chances that some memory swapping happened and that a fragment landed on disk, but again the chances are small and it was probably immediately overwritten. The same tiny chance applies for any equipment that the conversation traveled through.

To go further down that path you're really talking computer forensics - the stuff of computer detectives pulling apart hard drives to see what can be recovered even from deleted data. But as I said, it's unlikely there's anything to recover (Notenboon, 2004). There is no updated information about how to retrieve the messages from the messenger.

Instant messaging applications can provide strong evidence in certain cases (Farmer, 2008). Windows Messenger, MSN Messenger and Windows Live Messenger generally utilize any of the three following keys:

- HKEY-CURRENT-USER\software\Microsoft\MessengerService
- HKEY_CURRENT_USER\Software\Microsoft\MSNMessenger
- HKEY_LOCAL_MACHINE\Software\MICROSOFT\Messenger Service

TEST METHODOLOGY FOR COMPUTER FORENSIC TOOLS

The Computer Forensics Tool Testing project at the National Institute of Standards and Technology (NIST), an agency of the United States Department of Commerce, was designed to provide a measure of assurance for the software tools used by law enforcement in computer forensics investigations. This project should provide law enforcement personnel with a means of deciding whether the tools in consideration for use should be applied to the purposes required. These tools, in combination with the forensic processes and protocols allow investigators to collect digital evidence that can be introduced into a judicial process. This project is supported by agreements between NIST and the National Institute of Justice, the research and development organization of the U.S. Department of Justice, the Technical Support Working Group, and other agencies.

This paper provides a description of the general approach taken to develop the test methodology for computer forensic tools and the rationale behind this approach. The development of a test methodology was complicated by the lack of standards or specifications that describe what forensic tools should do and the need for these tools to survive the scrutiny of a judicial process. This methodology is based on NIST's extensive experience in testing, proven methodologies for conformance testing, and international standards and guidelines for testing.

BACKGROUND

Our approach for testing computer forensic tools is based on well-recognized methodologies for conformance testing and quality testing. ISO/IEC Guide 2, Standardization and Related Activities – General Vocabulary defines the basic terms and concepts that are used for international conformity assessment programs, testing and measurement, and general mutual recognition of testing results.

It defines conformity or conformance as the "fulfillment by a product, process or service of specified requirements." Conformance testing captures the technical description of a specification and measures whether an implementation faithfully implements the specification.

ISO/IEC 17025, General Requirements for the Competence of Testing and Calibration Laboratories provides recognized codes for good laboratory practice and criteria for quality assurance in testing laboratories. The guide serves as a reference not only for accrediting laboratories for independent testing, but also outlines requirements for quality testing so that users may have some confidence and assurance in the test results obtained. Our approach follows these principles so that our test methods meet the highest level of quality assurance possible.

Approach

The general approach for testing computer forensic tools can be summarized as follows:

1. Establish categories of forensic requirements,
2. Identify requirements for a specific category,
3. Develop test assertions based on requirements,
4. Develop test code for assertions,
5. Identify relevant test cases,
6. Develop testing procedures and method,
7. Report test results.

As suggested by ISO 17025 for cases where there is no standard test method, the results of each step will be made available for public review so that this process is an open, public process which incorporates and reflects the needs of a wide variety of law enforcement practitioners and suppliers of computer forensic tools.

FORENSIC REQUIREMENTS CATEGORIES

Forensic requirement categories are groupings of forensic functions that are determined by expert users. Grouping these forensic functions provides a smaller set of requirements that can be systematically approached by specialized forensic experts for testing. Narrowing the scope facilitates identifying the requirements for each functional grouping.

Forensic Requirements Specification

A forensic requirements specification prescribes the technical or functional requirements to be fulfilled by a product. Specifically, it identifies the requirements and features applicable to a category of forensic tools. A group of experts from federal, state, and local law enforcement organizations initially identify a list of requirements or specifications for the category of forensic functions.

However, the final requirements are based on consensus review from the communities that use the tools. These requirements will be used as the basis for developing test assertions and test cases that will be used to test the forensic tools.

Test Assertions

A test assertion is a statement of behavior, action, or condition that can be tested or measured. The test assertions bridge the gap between the narrative of the specification and the test cases. Each test assertion is an independent, complete, testable statement for a requirement in the specification. Each test assertion results in the realization of one or more test cases.

Test Cases

A test case specifies what is to be tested or one instance of what is to be tested. Test cases apply the assertions to the environments that are to be tested. Economic considerations limit the number of test cases that can be selected. Public review and the opinions of the group of experts from both computer science and forensic practitioners are used to narrow the number of test cases. In addition, the NIST Statistical Engineering Division is developing an independent experimental design to assist in the selection of test cases.

Test Method

The test method is a combination of the software used for testing and the procedures for completing the testing. If the test assertions and test cases are the 'what' that is tested, then the test method is 'how' the testing is to be accomplished. ISO/IEC 13210 "Information Technology – Test Methods for Measuring Conformance to POSIX" defines test method as "the software, procedures, or other means specified by a POSIX standard to measure conformance....Test methods are intended to provide a reasonable, practical assurance that the implementation conforms to the standard." Thus, our test method is the key to providing reasonable, practical assurance that the forensic tools will meet the requirements of the investigators.

ISO/IEC 17025 identifies the required documentation for any 'non-standard test method.' These 16 generic items were used to develop the requirements for forensic tool test method documentation.

a. *Appropriate identification*: Unique identifier which identifies the precise test case
b. *Scope*: The test method for which category of forensic requirements
c. *Description*: What product is being tested including version numbers
d. *Parameters*: The variables that are used to define the test, such as the size of the disks being tested, access, action, etc.
e. *Apparatus*: The testing environment or computers being used
f. *Reference standards*: The testing software or support software used
g. *Environmental conditions*: Where the testing was completed
h. *Description of the procedure*: Identifying and documenting which equipment is being used and any preparation of the equipment before testing checks to be made before testing begins including setup procedures identifying how the documentation will be kept, name of files, etc. identifying any procedures needed to protect the integrity of the test results
i. *Criteria for approval*: What should be the expected results
j. Data: What data will be captured, how the results will be analyzed, and presented

International guidelines state that there can be no assurance of the integrity, reliability, and correctness of test results unless the test method is valid. Section 5.4.5 of ISO/IEC 17025, "Validation of methods" states that:

- "Section 5.4.5.1 Validation is the confirmation by examination and the provision of objective evidence that the particular requirements for a specific intended use are fulfilled.
- Section 5.4.5.2 The laboratory shall validate non-standard methods, laboratory-designed/developed methods, ... to confirm that the methods are fit for the intended use. The validation shall be as extensive as is necessary to meet the needs of the given application or field of application. The laboratory shall record the results obtained, the procedure used for the validation, and a statement as to whether the method is fit for the intended use. ..."

When a test method is standardized, it has had the scrutiny of all participants who vote on the standard. When there is no standard, the review of experts in the field and any other interested parties is required. Our approach therefore is that all procedures used and all testing software will be made available for any interested party to use and judge. Most of the information, the specifications, the test assertions, test cases, and the support software will be made available on the Internet.

Test Results Reporting

Test results reporting is the way that the test results are presented. ISO 17025 states that "The results of each test, calibration, or series of tests or calibrations carried out by the laboratory shall be reported accurately, clearly, unambiguously and objectively, and in accordance with any specific instructions in the test or calibration methods. The results shall be reported, usually in a test report ..., and shall include all the information requested by the client and necessary for the interpretation of the test or calibration results and all information required by the method used ...."

This international standard gives specific information on what should be included in any test report "unless the laboratory has valid reasons for not doing so."

Thus, the forensic tool test results should include:

a. A title stating what product was tested;
b. Identification of the testing environment, i.e., where the tests were run;
c. Unique identifier for the test report; [identifier will be repeated on each page in order to ensure that the page is recognized as a part of the test report]
d. The name and address of the vendor;
e. Identification of the testing software used;
f. Unambiguous identification of the product tested including version, patches, etc.;
g. The test with the criteria for measurement;
h. The name(s), function(s) and signature(s) or equivalent identification of person(s) authorizing the test report;
i. Where appropriate and needed, opinions and interpretations;
j. Additional information which may be required by specific methods, clients, or groups of clients.

REPEATABILITY AND REPRODUCIBILITY

Test results must be repeatable and reproducible. The procedures of the test method in addition to the testing software ensure this. The basic concepts of these two conditions are defined in ISO 5725, "Accuracy (trueness and precision) of measurement methods and results."

- Repeatability: Precision under repeatability conditions.
  Repeatability conditions: Conditions where independent test results are obtained with the same method on identical test items in the same laboratory by the same operator using the same equipment within short intervals of time.
- Reproducibility: Precision under reproducibility conditions.
  Reproducibility conditions: Conditions where test results are obtained with the same method on identical test items in different laboratories with different operators using different equipment."

As applied to computer forensic testing, repeatability is defined as the ability to get the same test results on the same testing environment (same computer, disk, mode of operation, etc.) Reproducibility is defined as the ability to get the same test results on a different testing environment (different PC, hard disk, operator, etc.).

MANAGEMENT ISSUES AND DOCUMENTATION

Quality control and documentation of the testing process are key to insure repeatability and reproducibility of test results. Every step of the process will be documented at a level of detail that will provide enough information that the testing process, from setup to completion, could be repeated and analyzed. These records in addition to the testing hardware and testing software are controlled and protected to insure the accuracy and integrity of the process.

UTILITIES

INTERNET EXPLORER CACHE VIEW (IECACHEVIEW)

Is a small utility that reads the cache folder of Internet Explorer, and displays the list of all files currently stored in the cache. For each cache file, the following information is displayed: Filename, Content Type, URL, Last Accessed Time, Last Modified Time, Expiration Time, Number Of Hits, File Size, Folder Name, and full path of the cache filename. You can easily save the cache information into text/html/xml file, or copy the cache table to the clipboard and then paste it to another application, like an Excel Spreadsheet.

Advantages over the 'Temporary Internet Files' viewer of Windows

- IECacheView displays only the list of cache files, while the cache view of Windows displays a mix of cookies and cache files.
- IECacheView allows you to filter the cache files by file type (image, text, video, audio, or application).
- IECacheView allows you to view the cache files of another user or from another disk, while with the Windows viewer, you can only watch the cache of the current logged-on user.
- IECacheView displays some columns that are not displayed by the cache viewer of Windows: Content Type, Number of hits, Sub-folder name, and the full-path of the cached filename.
- With IECacheView, you can easily select the desired cache items, copy the information to the clipboard, and then paste it into an Excel or OpenOffice Spreadsheet.

Using IECacheView

- IECacheView doesn't require any installation process or additional DLL files. Just copy the executable file (IECacheView.exe) to any folder you like, and run it. After you run it, the main window displays the list of files currently stored in the cache of Internet Explorer for the current logged-on user.

If you want to view the cache of another user or from another instance of the operating system, simply use the 'Select Cache Folder' option (F9) to select the desired cache folder that you want to inspect.

- You can select one or more cache files from the list, and then export the list into text/html/xml file ('Save Selected Items' option), copy the URL list to the clipboard, copy the entire table of cache files, and then paste it to an Excel or to OpenOffice spreadsheet. You can also extract the actual files from the cache, and save them into another folder.

INTERNET EXPLORER HISTORY VIEWER (IEHISTORYVIEW)

Each time that you type a URL in the address bar or click on a link in Internet Explorer browser, the URL address is automatically added to the history index file. When you type a sequence of characters in the address bar, Internet Explorer automatically suggests to you all URLs that begin with the character sequence that you typed (unless the AutoComplete feature for Web addresses is turned off). However, Internet Explorer doesn't allow you to view and edit the entire URL list that it stores inside the history file.

This utility reads all information from the history file on your computer, and displays the list of all URLs that you have visited in the last few days. It also allows you to select one or more URL addresses, and then removes them from the history file or saves them into text, HTML or XML file. In addition, you are allowed to view the visited URL list of other user profiles on your computer, and even access the visited URL list on a remote computer, as long as you have permission to access the history folder

Using IEHistoryView

IEHistoryView is provided as standalone executable, so it doesn't require any installation process or additional DLLs. Just copy the executable (iehv.exe) to any folder you like, and run it. After you run it, the main window displays the list of all URLs stored in the history file of the current logged on user. If you want to view the history information of other users on your computer, choose the "Select User Profile" from the file menu, and select the desired user profile.

The Typed URLs List

Most URLs that you visit are saved in the History folder. However, Internet Explorer also saves the last 25 URLs that you typed in the following Registry key: HKEY_CURRENT_USER\Software\Microsoft\Internet Explorer\TypedURLs

By default, IEHistoryView doesn't display the URLs list from the Registry, but you can change this setting by choosing the "Display Typed URLs" option under the View menu. If this option is selected, the URLs list in the Registry are displayed, in addition to the list of all URLs stored in the History folder.

You can distinguish between the 2 types of URLs by looking at the 'Hits' column or at one of the date columns. As opposed to the URLs extracted from the History folder, The URLs from the Registry doesn't provide any additional information except the URL itself, so the other columns contain an 'N/A' string (Not Available).

Be aware that the typed URLs list can only be displayed for your local system. You cannot view the typed URLs list if choose to watch the history data from another computer or from another operating system.

About The History Folder

The location of the history folder is different from one operating system to another. On Windows 2000/XP, the History folder is located inside "Local Settings" folder of your user profile. For Example: C:\Documents and Settings\Administrator\Local Settings\History. The "Local Settings" folder is hidden by default, so you won't see this folder unless your system is configured to display hidden files and folders.

In most systems, IEHistoryView automatically identifies your current History folder and uses it as default. However, you can always select the History folder from another location by using the "Select History Folder" option under the File menu.

Be aware that when you watch the History folder from a Windows environment, it doesn't show you the real files inside this folder, but instead it displays the history shell extension that provides limited information about the sites you visited (Nirsoft, IEHistoryView, 2009)

MOZILLACACHEVIEW

Is a small utility that reads the cache folder of Firefox/Mozilla/Netscape Web browsers, and displays the list of all files currently stored in the cache. For each cache file, the following information is displayed: URL, Content type, File size, Last modified time, Last fetched time, Expiration time, Fetch count, Server name, and more. You can easily select one or more items from the cache list, and then extract the files to another folder, or copy the URLs list to the clipboard.

Mozilla Cache Folder Location

The cache folder of Mozilla Firefox is located under C:\Documents and Settings\[User Name]\Local Settings\Application Data\ Mozilla\ Firefox\ Profiles\[Profile Name]\Cache.

Using MozillaCacheView

MozillaCacheView doesn't require any installation process or additional DLL files. Just copy the executable file (MozillaCacheView.exe) to any folder you like, and run it. After you run it, the main window displays the list of files currently stored in the cache of the Mozilla/Firefox profile that you used in the last time. If you want to view the cache of another profile, simply use the 'Select Cache Folder' option (F9), and choose the desired cache folder.

You can select one or more cache files from the list, and then export the list into text/html/xml file ('Save Selected Items' option), copy the URL list to the clipboard, copy the entire table of cache files, and then paste it to an Excel or to OpenOffice spreadsheet. You can also extract the actual files from the cache, and save them into another folder (Nirsoft, MozillaIECacheView, 2009).

CHROMECACHEVIEW

Is a small utility that reads the cache folder of Google Chrome web browser, and displays the list of all files currently stored in the cache. For each cache file, the following information is displayed: URL, content type, file size, last accessed time, expiration time, server name, server response, and more.

Chrome Cache Folder location

The cache folder of Google Chrome is located under [User Profile]\Local Settings\Application Data\Google\Chrome\User Data\Default\Cache

Using ChromeCacheView

ChromeCacheView doesn't require any installation process or additional DLL files. Simply copy the executable file (ChromeCacheView.exe) to any folder you like, and run it. After you run it, the main window displays the list of files currently stored in the cache of the default Google Chrome user.

You can select one or more cache files from the list, and then export the list into text/html/xml file ('Save Selected Items' option), copy the URL list to the clipboard (Ctrl+U), copy the entire table of cache files (Ctrl+C), and then paste it to an Excel or to OpenOffice spreadsheet. You can also extract the actual files from the cache, and save them into another folder (Nirsoft, ChromeCacheView, 2009).

SQLITE

SQLite is an embeddable database system that uses flat files. It does not need to be started, stopped, configured, or managed like other SQL databases. It is lightweight, fast, and compact. And it works completely out of the box without any configuration. The entire database (definitions, tables, indices, and the data itself) is stored as a single cross-platform file on a host machine. This simple design is achieved by locking the entire database file during writing (Kristianto, 2009).

Advantage of SQLite

1. Zero-Configuration
2. Serverless
3. Single Database File
4. Stable Cross Platform Database File
5. Compact Size
6. Variable-length records
7. Free with Public Domain license

Also Google chrome and Mozilla Firefox use SQLite to storage the information.

Disadvantage of SQLite:

1. In most cases database size is restricted to 2GB
2. Not fully SQL92 compliant
3. Lock whole file while writing
4. No caching mechanism
5. Not very scalable

MICROSOFT LOG PARSER

Microsoft log parser queries everything with sql. It lets you slice and dice a variety of log file types using a common SQL-like syntax. It's an incredibly powerful concept, and the LogParser implementation doesn't disappoint. This architecture diagram from the LogParser documentation explains it better.

Log Parser operates as a kind of data pipeline. Into this pipe you can send information from IIS logs, Windows Event logs, Active Directory information, file system data, Registry data, Network Monitor traces, and so on. Once the data is in the pipe, you can process it using SQL statements; for example, to select certain portions of the data by a SELECT query. Then, as the processed data comes out of the pipeline, you can output it to text files, HTML files, Excel-style charts, or a SQL database table, or simply to the console as raw output.

Log parser consists of three components, which are: 1) input engine, 2) SQL query engine, and 3) output engine. The input engine and output engines are truly incredible and, combined, make this tool shine. When investigating network intrusions, you are faced with analyzing logs from many sources, none of them being compatible with the other. Log parser can accept most any common log format and output it into one of many formats of your choosing. When you are done, you can combine all your disparate logs into one common format for analysis.

At any point in the process you can subject your logs to a query so that you narrow down the data to that which is relevant. While many GUI tools are out there that provide filters, even those that allow the user to build custom filters can't compare with the power of writing a custom SQL query in Log Parser.

As an intrusion investigator / forensic examiner, you are tasked with mastering many tools to get your work done. It would be nice if we only had to master a couple of tools, but such will never be the case. We can however, limit the number of tools we have to use if we make careful selections. Whenever you can use one tool that will handle multiple tasks instead for multiple tools for the same number of tasks, that should be your tool of choice. Log parser fits this criteria as it can process and query all the common logs formats and can address your file system and your registry as well, including those of remote systems.

The best way to get to know this tool is to use it daily in the administration of your systems. You can create batch files to run your SQL queries against your logs, place them in your scheduler, and have critical log reports sitting on your desktop each day when you come to work. By getting to know this tool and its capabilities in this manner, you can apply those acquired skills to forensic applications of this tool. In the end, you'll have better management of your systems and have a forensic tool that you'll find new uses for with every case you process.

Using Log parser:

A basic SQL query must have, at a minimum, the following:

- SELECT
- FROM

To continue, you must have installed log parser. Open the command (cmd.exe) line interface in the root of the folder "C:\Program Files\Log Parser 2.2" wherein lies the executable "logparser.exe". From the command line interface, type in the following:

- logparser.exe -i:EVT -o:NAT "SELECT TimeGenerated, EventID FROM System

Another output feature of log parser is its "DATAGRID" output. Instead of dumping the query to a screen, you can send it to a GUI interface. To send it to a datagrid, enter the following:

- logparser.exe -i:EVT -o:DATAGRID "SELECT TimeGenerated, EventID, message FROM System" and you should see the following.

OUTLOOK REDEMPTION

Outlook Redemption works around limitations imposed by the Outlook Security Patch and Service Pack 2 of MS Office 98/2000 and Office 2002/2003/2007/2010 plus provides a number of objects and functions to work with properties and functionality not exposed through the Outlook object model

With Outlook Redemption you can:

- Make your code run unaffected by the Security Patch.
- Access properties not exposed by the Outlook Object Model (internet message headers, sender e-mail address and hundreds more properties)

- Display Address Book.
- Directly access RTF body of any Outlook item
- Import MSG, EML (RFC822) and TNEF files
- Export messages to the MSG, EML, TXT, HTML, TNEF, iCal and vCard formats.
- Access and manipulate Outlook accounts (Outlook 2002 and above, RDO library)
- Create, access and manipulate MAPI profiles and accounts (Profman library, see also RDOSession files)
- Access MAPI stores.
- Access Outlook nicknames.
- Manage Junk Mail settings
- Manage categories
- Directly access message attachments as strings or as arrays without saving them as files first

Redemption supports Outlook 98, 2000, 2002, 2003, 2007 and 2010-32 bit.

Redemption has two purposes:

- To allow programmatic control over Microsoft Outlook, without the interruption of the Object Model Guard.
- To give access to certain parts of the Microsoft Outlook Object Model that is not normally available. For example, Outlook users who want to be able to list out the internet headers, from the emails they receive will find that most of the header information is not accessible.

How Outlook Redemption works

Redemption is a regular COM library once registered on the system it is accessible to any programming language. Redemption uses Extended MAPI (which is not affected by the security patch since it is not accessible to the scripting languages) to duplicate the functionality blocked by the Security Patch. All safe item Redemption objects have an Item property which must be set to an Outlook item. Once set, you can access any properties and methods, both blocked and not blocked.

For the blocked properties and functions, Redemption objects completely bypass the Outlook object model and behave exactly like Outlook objects with no Security Patch applied For the properties and methods not blocked by the Security Patch, all calls are transparently forwarded to the Outlook object that you assign to the Item property. With this approach changes to your code are minimal: you only change the way you declare the objects, but not the rest of your code that actually accesses both blocked and not blocked properties and methods.

RELEVANT EVIDENCE

There is Inclusion and Exclusion Criteria weighted by the Investigators of the District Attorney's Office. Note that this is not an official statement, just an opinion based on hypothetical cases.

Inclusion Criteria

- There are more than one different activity e.g. visiting web site and sending an email and
- The time difference between activities should not be more than 10 minutes apart
- The more activities close in time, the more relevant is the evidence
- Show that user's history previously visited the same web site (very relevant)

Exclusion Criteria

- There is only one activity and no history of user's visiting the same web site
- More than one activity but more than 15 minutes apart and no history of user's visiting the same web site

6

Computer Investigation in Forensic Science

INTRODUCTION

For years the police have entered homes and offices, hauled away filing cabinets full of records, and searched them back at the police station for evidence. In Fourth Amendment terms, these actions are entry, seizure, and search, respectively, and usually require the police to obtain a warrant. Modern-day police can avoid some of these messy steps with the help of technology: They have tools that duplicate stored records and collect evidence of behavior, all from a distance and without the need for physical entry. These tools generate huge amounts of data that may be searched immediately or stored indefinitely for later analysis.

Meanwhile, it is unclear whether the Fourth Amendment's restrictions apply to these technologies: Are the acts of duplication and collection themselves seizure? Before the data is analyzed, has a search occurred? Today, tools can detect heat released from buildings, recreate images displayed on distant computer monitors, determine what is typed on a keyboard by listening to the distinct sounds of the key presses, and eavesdrop on Wi-Fi Internet communications traveling through the air. Handheld GPS units can monitor and store our movements around town, and web browsers keep detailed records of the websites we have visited. Tomorrow will surely bring new tools that are more invasive, easier to use, and able to work from greater distances (Ohm, 2005).

Computer forensics is a relatively new discipline to the courts and many of the existing laws used to prosecute computer-related crimes, legal precedents, and practices related to computer forensics are in a state of flux. New court rulings are issued that affect how computer forensics is applied. The important point for forensics investigators is that evidence must be collected in a way that is legally admissible in a court case (CERT, 2008).

Computer forensics involves obtaining and analyzing digital information for use as evidence in civil, criminal or administrative cases. Documents maintained on a computer are covered by different rules, depending on the nature of the documents. Many court cases in state and federal court have developed and clarified how rules apply to digital evidence. The Fourth Amendment of the US Constitution (and each state's constitution) protects everyone's right to be secure in their person, residence and property from search and seizure (Computer Forensics, 2008). Thus like any other crime scene, rules apply to obtaining search warrants to search and seize computers, computer files, and disks.

SEARCH AND SEIZURE

In computer forensics the search and seizure Fourth Amendment has played a fundamental roll. The fourth amendment states: The right of the people to be secure in their persons, houses, papers, and effects, against unreasonable searches and seizures, shall not be violated, and no Warrants shall issue, but upon probable cause, supported by Oath or affirmation, and particularly describing the place to be searched, and the persons or things to be seized (Wegman, 2004).

The Fourth Amendment is part of the Bill of Rights which guards against unreasonable searches and seizures. It was ratified as a response to the abuse of the writ of assistance which was a type of general search warrant in the American Revolution. It specified that any warrant must be judicially sanctioned for a search or an arrest in order for such a warrant to be considered reasonable. Warrants must be supported by probable cause and be limited in scope according to specific information supplied by a person.

It only applies to governmental actors and to criminal law (IST 432-Computer Forensic). An example would be if a warrant is issued for child porn on an individual's computer, but finds records of embezzlement, the embezzlement records could not be used in a court of law. The exception is if the police could justify obtaining a warrant to search the computer for records of embezzlement.

The Fourth Amendment interposes a magistrate as an impartial arbiter between the defendant and the police. The magistrate may issue a search warrant if the magistrate or judge is convinced that probable cause exists to support a belief that evidence of a crime is located at the premises. The officer must prepare an affidavit that describes the basis for probable cause and the affidavit must limit the area to be searched and evidence searched for.

The warrant thus gives the police only a limited right to violate a citizen's privacy. If the police exceed that limited right, or if a warrant is required, but the police have not first obtained one, then any evidence seized must be suppressed (U.S. Department of justice 2002). The issue of suppression driven by a determination of whether the Fourth Amendment has been correctly followed by the police is often the determining factor in criminal cases (Wegman, 2004).

Search warrants give only limited authority to the police to search. The search should be no more extensive than necessary as justified by probable cause. Thus, if the probable cause indicates that the contraband is located in a file on a CD, this would not justify seizing every computer and server on the premises. The extent of the search is tailored to the probable cause. If the police wish to seize a computer for analysis at a later time, the probable cause statement should demonstrate the impracticality or danger of examining the computer on the premises; hence the need to confiscate it and analyze it off-site (Wegman, 2004).

Another question facing law enforcement is when to notify the target of a search. Normally the target is notified at the time a physical search is made. However, the USA PATRIOT Act amended Title 18, Sec.3103a of the United States Code to permit delayed notification. Law enforcement may now delay notification of the target for up to 90 days, with another delay possible upon a showing of good cause.

In order to obtain authority for delayed notification, an investigator must show a need for the delay (IST 432- Computer Forensic). Reasons include danger to the life or safety of an individual, risk of flight from prosecution, witness or evidence tampering, or that immediate notice would seriously jeopardize the investigation.

Another legal issue in computer forensic cases is how much time the police may have to analyze a computer after seizing it. Federal Rule of Criminal Procedure 41 (c) (1) gives the police ten days after issuance of the warrant to serve it. But there is nothing in the Federal Rule of Criminal Procedure about how long the police may keep and analyze the computer. As a practical matter, the search of a computer in police custody should be done as quickly as possible. This is especially important if the computer is needed for the operation of a business (Wegman, 2004).

In the United States Supreme court case of Illinois v. Andreas, 463 U.S. 765 (1983), the Court held that a search warrant is not needed if the target does not have a reasonable expectation of privacy in the area searched. The loss of a reasonable expectation of privacy, and therefore the loss of Fourth Amendment protection is extremely important because much information is transmitted to networks and to the internet. If circumstances suggest the sender had no reasonable expectation of privacy, then no warrant is required by the police in order to obtain that information (Wegman, 2004). Examples would be blogs, website posts, and websites themselves. Public computers like library computers are not covered under the expectation of privacy.

No warrant is needed when the target consents to a search of his/her computer. No warrant is needed where a third party, such as a spouse, parent, employer or co-worker consents to the search, so long as the third party has equal control over the computer (USA Dept of Justice, 2009). An example would be if a married couple shared a computer in their home. The wife could consent to a search without the husband's consent and vice versa.

Agents should be especially careful about relying on consent as the basis for a search of a computer when they obtain consent for one reason, but then wish to conduct a search for another reason. In two recent cases, the Courts of Appeals suppressed images of child pornography found on computers after agents procured the defendant's consent to search his property for other evidence. In United States v. Turner, 169 F.3d 84 (1st Cir. 1999), detectives searching for physical evidence of an attempted sexual assault obtained written consent from the victim's neighbor to search the neighbor's "premises" and "personal property." Before the neighbor signed the consent form, the detectives discovered a large knife and blood stains in his apartment, and explained to him that they were looking for more evidence of the assault that the suspect might have left behind.

While several agents searched for physical evidence, one detective searched the contents of the neighbor's personal computer and discovered stored images of child pornography. The neighbor was charged with possessing child pornography. On interlocutory appeal, the First Circuit held that the search of the computer exceeded the scope of consent and suppressed the evidence. According to the Court, the detectives' statements that they were looking for signs of the assault limited the scope of consent to the kind of physical evidence that an intruder might have left behind.

By transforming the search for physical evidence into a search for computer files, the detective had exceeded the scope of consent. (Concluding that agents exceeded scope of consent by searching the computer after the defendant signed a broadly-worded written consent form, because agents told the defendant that they were looking for drugs and drug-related items rather than computer files containing child pornography) (USA Dept of Justice, 2009).

ELECTRONIC COMMUNICATIONS PRIVACY ACT (ECPA)

Congress has responded to the changing technological landscape. The most important federal statutes affecting computer forensics are the Electronic Communications Privacy ACT (ECPA), the Wiretap Statute, the Pen/Trap Statute and the USA PATRIOT Act(Wegman, 2004). Enacted in 1986, the Electronic Communications Privacy Act sets provisions for the access, use, disclosure, interception and privacy protections of electronic communications. Violations of the ECPA may result in criminal penalties and civil remedies, including punitive damage.

This act was written to expand the wiretapping provisions to wireless telephony (cellular) and email communications. The ECPA works to prohibit unauthorized interceptions or disclosure of electronic communications. According to the US code electronic communications "means any transfer of signs, signals, writing images, sounds, data or intelligence of any nature transmitted in whole or in part by a wire, radio, electromagnetic, photo electronic or photo optical system that affects interstate or foreign commerce, "thereby making much of the desired content of possible forensics searches out of reach (IST 432- Computer Forensic).

In more detail, the ECPA covers communications via pager, cellular and wireless telephony, browser requests, internet downloads, chat room traffic, voice mail and emails when transmitted by common carriers in interstate commerce. ECPA prohibits unlawful access and certain disclosures of communications contents. Additionally, the law prevents government entities from requiring disclosure of electronic communications from a provider without proper procedure (IST 432- Computer Forensic).

Computer forensics is affected a great deal by the ECPA. There are prohibitions in place against unlawful access to stored communications which include probing into RAM or disk drives for information in source or destination computer or during transit while the communication temporary intermediary storage such as on a server. Such a law may affect the searching of certain protected material; however, there are some exceptions under the ECPA. Currently the ECPA has not been updated to accommodate the Internet. Investigators have sought to use technologies which collect much more information than pen registers or trap and trace devices under the authority of this law. It should be strengthened to protect citizen's privacy in electronic communications (IST 432- Computer Forensic).

There are certain critical exceptions to ECPA. If the situation falls within an exception, the communications may be disclosed (18 U.S.C. & 2511(1) (18 U.S.C & 2702(b). Where an individual lacks an expectation of privacy law enforcement officers do not need a warrant to listen in. ECPA will not bar intercepting the communications in these instances. Where one has an expectation of privacy is not always clear. If I set up a rendezvous with an acquaintance in a secluded public park in the middle of the day, sitting on a solitary park bench, do we have an expectation of privacy?

According to DOJ, this inquiry embraces two discrete questions. First is whether the individual's conduct reflects "an actual (subjective) expectation of privacy" (IST 432- Computer Forensic). Secondly is whether the individual's subjective expectation of privacy is "one that society is prepared to recognize as 'reasonable'" (IST 432- Computer Forensic). In most cases, the difficulty of contesting a defendant's subjective expectation of privacy focuses the analysis on the objective aspect of the Katz test, i.e., whether the individual's expectation of privacy was reasonable (IST 432- Computer Forensic).

Courts foraying into cyberspace must shift their focus away from the two-prong Katz expectation of privacy test in order to preserve the values underlying the Fourth Amendment. In developing a new framework for expectation of privacy analysis in cyberspace, courts should focus on the historic context of the Fourth Amendment and the intent of its Framers. Government monitoring and analysis of click stream data is closely analogous to the general searches which the Founding Fathers sought to curtail in enacting the Fourth Amendment. Both types of searches are indiscriminate, exposing lawful activity along with contraband or unlawful action. Also both are incredibly intrusive, exposing intimate details about the lives of citizens to government scrutiny. A new rule needs to be established which recognizes that click stream data may be protected by the Fourth Amendment. Not because that the protection fits well with expectation of privacy analysis as developed by the Court in recent years, but rather because government click stream analysis is precisely the type of search the Framers intended to be subject to the Amendment's limitations (Winn, 2008).

Courts addressing this question should apply the normative analysis set forth by the Supreme Court in Smith v. Maryland instead of the rigid two-prong Katz test. The Court in Smith recognized that the two-prong Katz expectation of privacy test will sometimes provide an inadequate index of Fourth Amendment protection. In such situations, the Court explained, courts must undertake a normative inquiry to determine whether Fourth Amendment protection was appropriate. This normative inquiry asks a very simple question. Should an individual in a free and open society be forced to assume the risk that the government will monitor her as she engages in the activity at issue? Courts employing the normative inquiry "must evaluate the 'intrinsic character' of investigative practices with reference to the basic values underlying the Fourth Amendment" (Winn, 2008). Unlike the two-prong test, which assumes that society has already reached an objective conclusion about the proper amount of protection a particular activity deserves, the normative test acknowledges that society has not reached a consensus about the proper level of protection a certain activity warrants. In that case, the activity can be evaluated against constitutional norms (Winn, 2008).

Application of Smith's normative inquiry to click streams reveals that Internet users should retain an expectation of privacy in click streams, because this data is precisely the type of information the Framers sought to protect against arbitrary government intrusion. The Fourth Amendment was intended to limit government searches which held the potential to intrude into the intimate details of the private lives of citizens. Courts must recognize a legitimate expectation of privacy in the intimate records of our online activity in order to satisfy these constitutional norms (Winn, 2008).

The passage of the Fourth Amendment was the Framers' reaction to overly intrusive searches and seizures conducted by British and colonial authorities. Prior to the Amendment's passage, the colonists were plagued by the use of general warrants and writs of assistance which authorized law and customs enforcement officers to enter and search any building suspected of housing contraband (Winn, 2008). The searches conducted using these devices were broad and abusive, and occurred without particularized suspicion. The raids were led by executive officials with unlimited discretion (Winn, 2008). For example, the New Hampshire Council once allowed search warrants for "all houses, warehouses, and elsewhere in this Province", and the Pennsylvania Council once required a weapons search of "every house in Philadelphia" (Winn, 2008). Far from being isolated instances, such searches were widespread (Winn, 2008).

In response to these abuses, the Framers sought to limit the power of government actors to search or seize persons, houses, papers, and effects. The invasion the Framers sought to prohibit was not merely the physical intrusion upon a "person" or "house." Instead, "the amendment's opposition to unreasonable intrusion... sprang from a popular opposition to the surveillance and divulgement that intrusion made possible" (Winn, 2008). As one scholar explained, "The objectionable feature of general warrants was their indiscriminate character" (Winn, 2008). In addition to any contraband or unstamped goods that the generalized searches uncovered, the entirety of a person's private life was exposed to prying government eyes. This sort of indiscriminate search stripped the colonists of privacy without adequate justification, exposing them to the arbitrary and potentially despotic acts of government officials (Winn, 2008).

Monitoring and analysis of click streams by government officials is closely analogous to colonial general searches because it exposes the intimate lives of Web users, fails to discriminate between lawful and unlawful activity and grants enormous discretion to front-line executive officials. As with general searches of colonial homes, click stream searches will unnecessarily reveal private information to government view, even when this information pertains to lawful activity. For example, law enforcement agents monitoring click streams could learn that an outwardly heterosexual man spends time entertaining homosexual fantasies online in an adult chat room, or that a high-profile political leader used the Internet to reserve a spot in an addiction recovery center. While such conduct is certainly legal, it is also intensely private. Allowing government agents to expose the conduct of the innocent in order to pursue the guilty contradicts the purpose and intent of the Fourth Amendment (Winn, 2008).

On a more general level, the broad and arbitrary intrusion occasioned by a click stream search is contrary to "the most basic values underlying the Fourth Amendment" (Winn, 2008). Although the use of general warrants and writs of assistance undoubtedly motivated the Framers in drafting the Amendment, they did not intend its protection to be limited to the narrow purpose of outlawing general searches. Instead, the Amendment was intended to protect citizens against the type of arbitrary invasions by government into the lives of citizens which general searches typified. As one commentator explained:

While the history of the Fourth Amendment reveals many facets, one central aspect of that history is pervasive: controlling the discretion of government officials to invade the privacy and security of citizens, whether that discretion be directed toward the homes and offices of political dissentients, illegal smugglers, or ordinary criminals.(Winn, 2008) Similarly, the Supreme Court has repeatedly recognized that the harm the Fourth Amendment seeks to prevent is not the tangible invasion of one's person, papers, effects, or home, but rather the intangible invasion upon the sanctity and privacy of those objects occasioned by an unreasonable search or seizure (Winn, 2008). The indiscriminate nature of click stream searches illustrates their incompatibility with the values upon which the Fourth Amendment was based. As one scholar argued:

The first problem with indiscriminate searches is that they expose people and their possessions to interferences by government when there is no good reason to do so. The concern here is against unjustified searches and seizures: it rests upon the principle thate very citizen is entitled to security of his person and property unless and until an adequate justification for disturbing that security is shown. The second problem is that indiscriminate searches and seizures are conducted at the discretion of executive officials, who may act despotically and capriciously in the exercise of the power to search and seize. This latter concern runs against arbitrary searches and seizures; it condemns the petty tyranny of unregulated rummages. (Winn, 2008)

Absent an expectation of privacy in click stream data, law enforcement agents will be free to rummage through our online lives, revealing intensely private conduct. The Founding Fathers found the ability to conduct such arbitrary and suspicion without reason searches to be one of the most offensive aspects of general warrants and writs of assistance. This was clearly intended such searches to be illegal. Allowing such intrusions into private cyberspace activity merely because an outdated expectation of privacy test would find assumption of risk or the absence of a subjective expectation of privacy in click stream data does intense violence to the values underlying both the Fourth Amendment and a free society. Yet this is exactly the result that will be reached if courts continue to cling to Katz's two part test.

Once an expectation of privacy is established in click stream data, traditional Fourth Amendment principles regulating the reasonableness of searches and seizures can easily be applied. The traditional test of reasonableness, which balances the nature and quality of the intrusion upon an individual's Fourth Amendment interests against the importance of the governmental interests alleged to justify the intrusion, is perfectly suited for cyberspace. This test allows courts to protect against overly extensive and indiscriminate intrusion into online lives while also acknowledging that a sufficiently compelling governmental interest may justify such searches. This is the question that should be getting asked in every click stream search. However, it will never be asked until courts loosen their vise grip on the two-prong Katz test and decide that Internet users should retain a legitimate expectation of privacy in click stream data (Winn, 2008).

ECPA is a highly nuanced example of public policy. Congress felt that information stored on a network deserved varying levels of privacy protection, depending on how important or sensitive the information was. Accordingly, in Title 18, section 2703 of the U.S Code ECPA created five categories of sensitivity. The more sensitive the category, the greater the justification the government must show in order to obtain the information from a third party (usually the system administrator). The most sensitive information consists of the content of un-retrieved communications such as email that has resided in electronic storage for 180 days or less. After one hundred eighty days the information is considered "stale" and not deserving of the top category of protection, so does not require a full search warrant for access (Bui, Enyeart, Luong., 2003). The least sensitive category includes only basic information such as the name of the subscriber and how bills are paid. To obtain that information, the government needs only an administrative subpoena. An administrative subpoena can be issued by a government agency on its own, without prior approval by a court. For example, the FBI could issue an administrative subpoena for good cause. That subpoena could later be challenged, and if a court later decided that good cause did not exist then information obtained under that subpoena would be suppressed (Bui et all. 2003).

WIRETAP STATUTE

The Wiretap Statute (Title III) was amended 2001. While ECPA regulates government access to stored computer information in the hands of third parties, the Wiretap statute deals with direct surveillance or real time interception of electronic communications by government agents. Wiretaps most commonly affect telephone conversations (IST 432- Computer Forensic). Wiretap requires special judicial and executive authorization. An application for interception may not be filed unless it is first authorized by the attorney general or a specially designated deputy or assistant. The application must identify the officer authorizing the application. Attached to the government application should be the authorization, as well as copies of the attorney general's designations of those Department of Justice officials who have been authorized to approve wiretaps. Unlike traditional search warrants, a federal magistrate judge is not authorized to issue a wiretap. Only a federal district or circuit court judge may issue a wiretap. The application must contain a full and complete statement of the facts and circumstances relied upon to support a belief that an interception order should issue. The issuing judge must determine that there exists probable cause to believe that particular communications concerning the alleged offenses will be obtained through interceptions of communications. Before an interception order may issue, the judge must find:

- Probable cause for belief that a particular enumerated offense is being committed.

- Probable cause for belief that particular communications concerning that offense will be obtained through interception.

Besides a sufficient factual predicate like probable cause, the Fourth Amendment requires that every search be reasonable. As with any other search, whether an electronic search is reasonable depends upon balancing the degree of intrusion against the need for it. Thus, because an order to surreptitiously intercept private conversations is such an intrusive search, the application for interception must show more than mere probable cause, it must also show necessity. The application must contain a full and complete statement as to whether other investigative procedures have been tried and failed, or the reasons why such procedures reasonably appear to be unlikely to succeed or to be too dangerous if tried. The issuing judge must find that normal investigative procedures have been tried and failed or reasonably appear unlikely.

A wiretap may issue only for particular crimes. The application must contain a full and complete statement regarding the details as to the particular offense that has been, is being, or is about to be committed. The issuing judge must find probable cause to believe those particular crimes are being committed, have been committed, or are about to be committed by an individual. The identities of persons to be intercepted must be particularly described in the application and order. The nature and location of the communication facilities to be intercepted must be particularly set forth in the application and order. The application must contain a particular description of the type of communications sought to be intercepted.

The issuing judge must determine that there exists probable cause to believe that particular communications concerning the alleged offenses will be obtained through interceptions of communications. The application and order must set forth either that interception will cease after the particular communication sought is first intercepted or that interception will continue for a particular time period. Requirements of the Fourth Amendment is to prevent the execution of the overbroad general warrant abhorred by the colonists which results in a general, exploratory rummaging in a person's belongings. Given the intrusive nature of an interception order, the Wiretap Act incorporates a number of provisions which circumscribe the scope of the warrant and guard against law enforcement officers generally rummaging through phone calls.

The order for interception must contain a provision requiring the officers to execute the order in a manner whereby the interception of calls not particularly described and not otherwise subject to interception will be minimized. Similarly, no order may be entered authorizing interception for a period of time longer than necessary to achieve the objective, but in no event shall the authorization exceed thirty days. Three U.S. federal statutes govern the interception, accessing, use, disclosure and privacy protections of electronic and wire communications. The U.S. Electronic Communications Privacy Act (ECPA, 18 U.S.C. §§ 2701-2712) of 1986 covers stored communications. Real-time interception, as in wireless networks, is covered by the Pen/Trap Statute, 18 U.S.C. §§ 3121-3127, centered in addressing information (like 802.11 protocol headers), and by the Wiretap Statute ("Title III"), 18 U.S.C. §§ 2510-2522, centered in the contents of communication.

PEN/TRAP STATUTE

The Pen/Trap Statute was amended in 2001. The Pen/Trap Statute, 18 United Sates Code Sec. 3121-3127, provides for a less intrusive form of government surveillance than wiretap statue; it authorizes the installation of pen registers and trap and trace devices. A pen register records only dialing, routing and addressing information regarding outgoing electronic communications. Electronic communications include telephone, computer, telegraph and telex communications. A trap can trace device records the same information regarding incoming electronic communications.

The significant fact regarding both becomes that the content of communications is not recorded. Only information such as telephone numbers of incoming and outgoing calls is recorded. Because these devices record less sensitive private information the legal burden upon the government is significantly less than with a wiretap. Court orders for a pen/trap device require only a statement by the investigator that is the investigator's belief that the information likely to be obtained is relevant to a criminal investigation. A recitation of probable cause is not necessary, nor is it necessary to attest to the many other requirements necessary to obtain a wiretap order or a search warrant (Wegman, 2004).

To obtain an order, applicants must identify themselves, identify the law enforcement agency conducting the investigation and then certify their belief that the information likely to be obtained is relevant to an ongoing criminal investigation being conducted by the agency. The law prohibits unlawful monitoring and disclosure of the content of communications. It also mandates law enforcement to follow proper procedures to review electronic communications, such as the search and seizure electronic evidence procedures detailed in the "Searching and Seizing Computers and Obtaining Electronic Evidence in Criminal Investigations" document by the US DOJ, specifically sections III and IV, focused on electronic communications and surveillance.

USA PATRIOT ACT

On October 26, 2001 President Bush signed the Uniting and Strengthening America by Providing Appropriate Tools Required to Intercept and Obstruct Terrorism Act (USA PATRIOT Act). This Act was overwhelmingly passed by Congress shortly after the events of September 11, 2001. It expands the government's investigative power. This Act has become very controversial drawing criticism from both Conservatives and Liberals who question whether the Act goes too far.

Perhaps the most controversial provision of the PATRIOT Act is the so-called "sneak and peek" authority conveyed in Section 213 of the Act. This Section provides delayed notification to the targets of searches. The Act modifies the U.S. Criminal Code, Title 18, Sections 3103a and 2705. These modifications allow the government to delay notification of physical searches for up to ninety days. Extensions may be given for good cause. However, the delayed notification provision is restricted to cases where the government demonstrates an urgent need for delay, including situations where the life or physical safety of an individual is in jeopardy or to avoid the destruction of evidence. Excerpts of Section 2705 are reproduced in Appendix A.

Delayed notification is not an entirely new element in federal criminal law. It is the norm in wiretap cases, as noted above, has been used and upheld in the seminal U.S. Supreme Court case of Dalia v. U.S. in 1979. In that case federal investigators entered a home, searched and implanted a hidden microphone pursuant to a search warrant. Notice was delayed until the surveillance ended. What is new about the PATRIOT Act is that it provides for delayed notification in ordinary physical searches. In the past delayed notification has been used only in connection with electronic surveillance (Waxman, 2004).

The Act also makes it easier for law enforcement to install an electronic surveillance device. Formerly, a wiretap order or pen register order had to be obtained in the jurisdiction in which the device was to be installed. Internet communications typically involve Internet service providers (ISP) located in many jurisdictions. Sections 216 and 220 allow devices to be installed anywhere in the U.S.A.

The Act is of particular importance to computer forensic investigators and providers of information to the government. It gives immunity from civil lawsuits to any person who provides technical or other assistance in obtaining electronic information pursuant to a court order or valid request for emergency assistance.

- The PATRIOT Act contains numerous other provisions expanding the scope of forensic investigations. However, it also contains a sunset provision. Under this provision the Act will terminate on December 31, 2005, unless Congress votes to extend it. The sunset provision does not apply to the entire Act. Significant sections, including those authorizing delayed notification, national wiretap and pen register orders will not sunset automatically. The Obama administration supported the revisions to the law as approved by the committee and on Feb 2010, it was extended for one more year. The three sections of the Patriot act that would stay in force:

- Authorize court approved roving wiretaps that permit surveillance on multiple phone
- Allow court approved seizure of records and property in anti-terrorism operations
- Permit surveillance against a so called lone wolf, a non-US citizen engaged in terrorism who may not be part of a recognized terrorist group.

Computer forensics is specifically supported by the PATRIOT Act. Section 816 authorizes the expenditure of $50 million for the creation and support of regional computer forensic laboratories. These laboratories will conduct investigations and also train investigators (wegman, 2004).

COLORADO HOUSE BILL AMENDMENT

Oral advisement and written consent prior to search of a vehicle or a person during a police contact.

- Prior to conducting a consensual search of a person who is not under arrest, the person's effects or a vehicle, a peace officer shall comply with paragraph.
- A peace officer may conduct a consensual search only after articulating the following factors to, and subsequently receiving consent from, the person subject to the search of the person with the apparent or actual authority to provide permission to search the vehicle or effects. The factors are:
  - The person is being asked to voluntarily consent to a search; and
  - The person has the right to refuse the request to search
- After providing the advisement required in paragraph of this subsection(1), a peace officer may conduct the requested search only if the person subject to the search voluntarily provides verbal or written consent. Other evidence of knowing and voluntary consent may be acceptable, if the person is unable to provide written or verbal consent.
- A peace officer providing the advisement required pursuant to subsection (1) of this section need not provided a specific recitation of the advisement, substantial compliance with the substance of the factors is sufficient to comply with the requirement
- If a defendant moves to suppress any evidence obtained in the course of the search, the court shall consider the failure to comply with the requirements of this section as a factor in determining the voluntariness of the consent.
- This section shall not apply to a search conducted pursuant to section 16-3-103 C.R.S. a valid search incident to or subsequent to a lawful arrest, or to a search for which there is a legal basis other than voluntary consent. This shall include, but not be limited to, a search

in a correctional facility or on correctional facility property, a detention facility, county detention facility, custody facility, juvenile correctional facility of any mental health institute or mental health facility operated by or under a contract with the department of human services, a community corrections facility or a jail or a search of a person subject to probation or parole by a community supervision or parole officer when the person has consented to search as a term and condition of any probation or parole.

ROLES OF COMPUTER FORENSIC

The issue most related to computer forensics has to do with wire-tapping and warrant gathering. The bill changes the ability of the government to delay the notification of a warrant by up to ninety days after the search. In the past, it had been possible to delay notification when doing surveillance such as wiretaps, since it would be pointless to listen in on a conversation when the parties involved know of the surveillance. This was upheld in the case Dalia v. U.S., where a wiretap was used and notification was delayed. The change in the PATRIOT Act, however, extends this ability to actual physical searches, including the search of computers. This can theoretically be very helpful, as it is can be an easy process to remove data from a hard disk, but combined with the ability of not needing a warrant in terrorist matters can be a very infringing ability.

As alluded to the USA PATRIOT Act also allows investigators to act prior to actually obtaining a warrant, as long as the individual involved personally feels that a threat is inherent. It also prevents third parties who aid in the surveillance from being liable in a civil case. This, however, can be conflicting. There could theoretically be times where a government agent feels there is a threat and elicits the help of another, but then the third party might not be protected if a warrant is not granted in the future. This is definitely an issue that is relevant to computer forensics, as an ISP may grant access to a government official, only to then be held liable for granting that access in the future (IST 432- Computer Forensic).

Computer forensics is about investigating digital evidence related to criminal or suspicious behavior where computers or computer and related equipment may or may not be the targets. This process of identifying, preserving, analyzing and presenting digital evidence in a legally acceptable manner is not much different from traditional forensic science. The only difference is that the former focuses on digital evidence, whereas the latter focuses on physical evidence. Casey defines digital evidence as:

Any data stored or transmitted using a computer that support or refute a theory of how an offence occurred or that address critical elements of the offence such as intent or alibi. Digital evidence includes computer generated records such as outputs of computer programs and computer-stored records such as email messages. It is important to criminal investigations because it can be used as proof of crime, connection or alibi. However, handling digital evidence is challenging because the evidence can be easily hidden, manipulated or altered. Moreover, it is difficult to attribute certain computer activities to an individual especially in a multi-access environment. Similar to physical evidence digital evidence provides only a partial view of what may have happened.

The field of computer forensics has become a critical part of legal systems throughout the world. As early as 2002 the FBI stated that fifty percent of the cases the FBI now opens involve a computer (Reyes, Wiles., 2007). However, the accuracy of the methods and therefore the extent to which forensic data should be admissible is not yet well understood. Therefore, it is not yet safe to make the kinds of claims about computer forensics that can be made about other kinds of forensic evidence that has been studied more completely, such as DNA analysis. The accuracy of DNA analysis is well understood by experts, and the results have been transformational both in current and previous court cases.

DNA evidence has been instrumental in convicting criminals, and clearing people who have been wrongly convicted and imprisoned. DNA evidence condenses to a single number (alleles) with a very small, and will depend on probability of error. On the other hand, computer forensic evidence has matured without foundational research to identify broad scientific standards, and without underlying science to support its use as evidence. Another key difference between DNA and computer forensic data is that DNA evidence takes the form of tangible physical objects created by physical events. Contrast these to computer objects that are created in a virtual world by computer events (IST 432- Computer Forensic).

The technology of computers and other digital devices is evolving at an exponential pace. Existing laws and statutes simply cannot keep up with the rate of change. Therefore, when statutes or regulations do not exist, case law is used. Case law allows legal counsel to use previous case similar to the current one because the laws do not yet exist. Each new case is evaluated on its own merit and issues.

COMPUTER FORENSIC INVESTIGATION

When conducting a computer investigation for potential criminal violations of the law the legal processes one follows depends on local custom, legislative standards and rules of evidence.

In general, however, a criminal case follows three stages:

- The complaint,
- The investigation,
- The prosecution.

A criminal case begins when someone finds evidence of an illegal act or witnesses an illegal act. The witness or victim makes a complaint to the police.

Based on the incident or crime, the complainant makes allegations, an accusation or supposition of fact that a crime has been committed. A police officer interviews the complainant and writes a report about the crime. The police department processes the report and the department's upper management decides to start an investigation, or log the information into a police blotter. The police blotter provides a record of clues to crimes that have been committed previously. Criminals often repeat actions in their illegal activities, and these habits can be discovered by examining police blotters. This historical knowledge is useful when conducting investigation especially in high technology crimes (Nelson et all., 2008).

The investigator assigned to the case should be a specialist in retrieving digital evidence or computer forensic expert. After the investigator builds a case the information is turned over to the prosecutor. When conducting a computer investigation for a business, remember that businesses must continue with minimal interruption from an investigation. Because businesses usually focus on continuing their usual operations and making profits, many in a private corporate environment consider an investigation and apprehension of a suspect secondary to stopping the violation and minimizing damage or loss to the business.

Law enforcement officers often find computers and computer components as they are investigating crimes, gathering other evidence or making arrests. With digital evidence, it is important to realize how easily key data such as last access date, can be altered by an overeager investigator who is first at the scene. The U.S Department of Justice (DOJ) lays out a procedure in a manual that reviews proper acquisition of electronic evidence.

The authenticity and integrity of the evidence examined will be of critical importance. The first step is to establish a chain of custody policy for your organization. The goal of the policy is to ensure that each piece of evidence collected is accountable to an individual until it is either returned to its original owner or disposed of (Reyes, Wiles., 2007).

Computing investigations demand that you adjust procedures to suit the case. For example, if the evidence for a case includes an entire computer system and associated storage media, such as floppy disks, cartridges, tapes and thumb drives, an investigator must be flexible when accounting for the entire item. Some evidence is small enough to fit into an evidence bag. Other items, such as the monitor and printer are too large. To secure and catalog the evidence contained in large computer components an investigator can use large evidence bags, tape, tags, labels and other products available from police supply.

Be cautious when handling a computer component to avoid damaging the components, or coming into contact with static electricity which can destroy digital data. For this reason, an investigator needs to use antistatic bags when collecting computer evidence. An investigator might consider using an antistatic pad with an attached wrist strap as well. Both help prevent damage to computer evidence. Computer components require specific temperature and humidity ranges also. If it is too cold, hot, or wet, computer components and magnetic media can be damaged. Even heated car seats can damage digital media. Placing a computer on top of a two-way car radio in the trunk can damaged magnetic media. When collecting computer evidence, an investigator must have a safe environment for transporting and storing it until a secure evidence container is available (Nelson et all., 2008).

In traditional, old fashioned cases, a detective would receive information from a reliable informant that contraband, for example drugs, are located at a premises. The detective would prepare a statement describing the informant's reliability and that the informant had recently observed drugs at the premises. The detective would take the affidavit to a judge, who would determine whether probable cause existed. If that determination was positive, the judge would sign the search warrant authorizing the detective to search for and seize a specific type and quantity of drugs at that premises.

The detective would then go to the location and execute the warrant (Skibell 2003). However, in computer forensics cases there is added complexity. The contraband might consist of child pornography, or records of drug sales. This information might be located on a laptop computer, but it might also be located on a network server in another state or in a foreign country. The information might be located on a hard drive, a diskette or a CD. The contraband information might be very difficult to recognize, it could be encrypted, misleadingly titled or buried among a large number of innocent files (Weigman, 2004). It could take considerable time to identify the contraband.

As noted above, a search warrant gives only limited authority to the police to search. The search should be no more extensive than necessary, as justified by probable cause. Thus, if the probable cause indicates that the contraband is located in a file on a CD, this would not justify seizing every computer and server on the premises (Brenner 2002). The extent of the search is tailored to the extent of the probable cause. If the police wish to seize a computer and analyze it at a later time, the probable cause statement should demonstrate the impracticality or danger of examining the computer on the premises hence the need to confiscate it and analyze it off-site.

7

Forensic Medicine Use in Identification, Individualization and Classification of Physical Evidence

INTRODUCTION

Forensic science is the application of science pertaining to the law and requires the complementary interaction of a wide range of scientific specialties and disciplines. The 1st aspect of applied forensic science begins with the identification, individualization or classification of physical evidence. For some types of evidence, identification or individualization may only be possible after conducting chemical or scientific tests. Types of evidence that require testing to insure accurate identification include blood stains, body fluids, drugs, arson accelerants and other chemicals. The identification of unknown substances or objects may be achieved by comparing their characteristics with those of known standards, previously established criteria or database information.

In the forensic examination of fibers and hairs, determination of fiber type, form, dye composition, elucidation of color, species or anatomical origins utilize class characteristics for such identification. The ultimate goal of the identification process in forensic science is individualization: to say that a particular piece of evidence originates from a specific locus, scene or person. In reality, few types of evidence can be unequivocally individualized like fingerprint and deoxyribonucleic acid evidence. All other types of evidence, if appropriately identified, can be said to be consistent with originating from a particular source, site or individual.

Many of the classical databases, techniques and tests now routinely used in forensic identification have been constructed, developed and refined in a process which, for some disciplines, has taken centuries, and for more recent technologies, decades. To depict the historical development and growth of forensic sciences, this narrative will broadly address documented instances of the organizational and developmental aspects of several disciplines and specialties within the profession, outlining the progression from basic conception to the application of some methodologies and techniques used today. The evolution of the field originates with the worldwide development of a number of different specialty or subspecialty scientific disciplines, with the rate of advance or development of each discipline being extremely variable, due to complexity, geographical location or availability of financial resources and technology.

FORENSIC MEDICINE AND TOXICOLOGY

The earliest documented forensic specialty could probably be considered to be forensic medicine, also known as legal medicine or medical jurisprudence, and is the application of medicine and medical science to answer legal problems. Records indicate that the first documented dissertation on forensic medicine was written in China in the 6th century by an individual named Hsu Chich'Ts' si. Although this work was apparently lost, a 2nd Chinese manuscript still exists.

Completed in 1247, the Hsi Duan Yu (The Washing Away of Wrongs) provides what is thought to be the first alliance of medicine and law and supplies specifications on distinguishing death by drowning or strangulation from death by natural causes. Further records suggest that legal medicine really began to prosper sometime in the 6th century, heralding the development of modern day science and medicine. It was not until around the end of the eighteenth century that the first appearance of legal medicine experts in the courtroom was documented.

The development of modern-day chemistry is considered to have begun at the end of the 18th century, paving the way for the development of modern toxicology. Closely related to forensic medicine, forensic toxicology centers on the determination of toxic substances in human tissues and organs and the subsequent determination of the role any toxic agents may have in contributing to or causing death. In general, forensic toxicology cases entail some form of drug or alcohol abuse. One of the most influential people in the development of toxicology was Mathieu Orfila. Orfila moved to France in 1807 and eventually became dean of the medical school in Paris.

He is considered to be one of the first experts to provide reliable scientific evidence in a criminal trial, while his academic studies provided toxicology a firm niche within the developing sciences. Orfila and associates are also credited with developing the 1st chemical test to detect arsenic, the poison of choice at that time.

However, it was James Marsh, a Scottish chemist, who was considered to be the first to provide toxicological evidence supporting arsenic detection for a legal trial, around 1836. The Lafarge case, also conducted by Orfila, is documented as the first case in which the defense called an opposing expert, Francois Vincent Raspail, in an attempt to refute the scientific evidence of the prosecution's expert witness.

Personal Identification

Personal identification plays a mammoth role in forensic and criminal investigations, and many sub-specialties in personal identification are still being developed and more frequently utilized. Forensic odontology uses dental records to facilitate human identification. This type of identification process is often called for in the identification of unrecognizable bodies, after mass disasters, and in the identification and comparison of bite marks.

Also facilitating personal identification is the use of forensic anthropology. This discipline, mainly concerned with bodily or skeletal remains, has, over recent years, constructed a number of databases cataloguing differences in physical structure of the body as a function of sex and race. It was Alphonse Bertillon who first developed the most significant and earliest personal identification system, at the end of the 19th century.

Bertil-lon's system of anthropometry, which involved a series of body and facial measurements, was developed to establish a process for use in the identification of habitual criminals. The system, also known as bertil-lonage, was devised in Paris and applied worldwide. Nowadays, it is considered that this system of identification cannot be utilized to uniquely identify one person as being distinct from all others.

Although the concept of anthropometry was considered theoretically possible, in practice, inconsistency in measurement techniques decreased the discriminating ability of the system. Despite this, Bertillon's system is still considered to have made a significant contribution to the development of forensic science, representing one of the first examples of the use of individualization for criminal investigation. Anthropometric systems were replaced as the science of fingerprinting was developed and the value of fingerprinting in personal identification and individualization was acknowledged.

Fingerprints

Fingerprint examination is probably the most used of personal identification techniques. Much of the work involves the study and classification of fingerprints, the development of latent prints and the comparison of known and unknown fingerprints. Despite the fact that Bertillon opposed the introduction of fingerprinting, he was the first person in Europe to use latent prints to solve a criminal case, as well as being among the first to use systematic crime scene photography.

The development of fingerprinting for personal identification was originally the initiative of William Herschel, a British civil servant residing in India in 1877, who first publicly proposed that fingerprints might be useful in the identification of criminals.

However, it was Henry Faulds, a Scottish physician working in Japan around the same time, who recognized the value of latent prints from a crime scene. The use of fingerprints in identifying a criminal was subsequently published in the journal Nature, and Faulds continued on to use fingerprints to identify a Tokyo burglar in 1880. As a result of his own studies and fingerprint pattern types described by Francis Galton, Juan Vucetich, an Argentinean police official, set up his own fingerprint classification system in 1891.

In 1892, Galton published Fingerprints, the first text on the use of fingerprints in personal identification and crime investigation, and devised the first scientific methods for the classification of fingerprint patterns. Several years later, Sir Edward Richard Henry, another British civil servant in India, continued to devise a fingerprint classification scheme, cataloguing sets of fingerprints that could be retrieved for identification purposes; he subsequently developed the fingerprint classification system that would replace anthropometry in Europe and North America.

This system was implemented in Europe when Henry was appointed head of Scotland Yard in the early 1900s. At the same time, in the United States, DeForrest initiated the first systematic use of fingerprints in personal identification. More recently, the Federal Bureau of Investigation's (FBI's) Integrated Automated Fingerprint Identification System (IAFIS) was initiated as a means of providing identification services to US law enforcement communities.

Criminalistics

The discipline of criminalistics encompasses all areas of trace evidence, such as soil, glass, hairs, fibers, blood and other body fluids, including, saliva, sweat, semen and vitreous humor. Criminalistics also includes arson, explosives, drug identification and investigation, interpretation of pattern and imprint evidence, and is by far the broadest of the disciplines of forensic science.

As well as conducting toxicology studies, Orfila was also responsible for the inception of early studies on blood and semen identification. Some of Orfila's earlier work laid the foundation for the development of some of today's routine tests. As early as 1827, Orfila discovered that extracting intact spermatozoa from seminal stains was extremely difficult and developed a series of chemical tests for semen identification. Following subsequent studies, other scientists established that microscopical examination of semen stains provided the most useful forensic information.

In 1891, Hans Gross of Austria first promoted the concept of criminalistics by applying scientific knowledge and methods to facilitate the interpretation and analysis of physical evidence for crime investigation. Gross, also a legal expert, saw the function of crim-inalistics as the complementary collaboration of various forensic specialists with appropriate scientific backgrounds for the examination and interpretation of physical evidence. Gross was also responsible for the emancipation of a journal dedicated to forensic sciences. In France, Victor Balthazard and Edmond Locard conducted similar work.

Balthazard, medical examiner for the city of Paris, also performed studies on probability models for fingerprints, bullet comparison, animal hairs and blood spatter patterns. In 1910, Locard set up the first police crime laboratory in Europe, in Lyon, the location of today's Interpol, and is recognized for the inception of the Locard exchange principle: «Every contact leaves a trace», the basic tenet of forensic science. Also during the 1920s, Luke May was the 1st to develop and apply striation analysis for toolmark comparison, incorporating statistical analysis into his methodology.

In the latter part of the 1920s, Los Angeles Chief of Police August Vollmer, of the Los Angeles Country Sheriff's department, developed the 1st United States police crime laboratory. The FBI crime laboratory was subsequently established in 1932, and 5 years later Paul Kirk set up the 1st academic criminalistics program in the United States at the University of California. In 1950, the American Academy of Forensic Science was formed, with the subsequent institution of the Journal of Forensic Science.

BALLISTICS AND FIREARM IDENTIFICATION

The specialization of ballistics and firearm examination relates to firearm identification, comparison of bullet markings and rifling, the identification of projectiles, cartridge and shell cases, and determination of bullet trajectories and damage. The historical development of these specialties begins around 1835 when Henry Goddard, working for Scotland Yard, first used bullet comparisons to provide evidence leading to the arrest of a murderer.

The comparison revealed a flaw in the bullet that could be traced back to a flaw in the original bullet mold. Nearly 55 years later, Dr Alexandre Lacassagne, professor of forensic medicine at the University of Lyon, was the 1st to associate bullets with a specific gun, by comparing striations etched on the bullet from a murder victim and striations from the gun barrel from which the bullet was fired.

Lacassagne was also one of the 1st to study the nature of blood spatter patterns. The discipline further developed around 1900, when Paul Jesrich took photomicrographs of bullets for comparison, and illustrated the use of this technique for the possible individualization of minutiae.

It was not until 1913 that Victor Balthazard published the 1st scientific article on the significance of bullet markings and their relation to bullet individualization.

Further developments occurred during the 1920s, as Charles Waite first cataloged information regarding weapon manufacture, and, together with Goddard, Gravelle and Fisher, further developed the technique of comparison microscopy for bullet comparison.

Questioned documents

Another division of criminalistics is questioned document examination, which entails the comparison and interpretation of handwriting; materials generated from typing, printing, facsimiles and photocopying; and the analysis and aging of paper, inks and materials used to produce documents. Such techniques were used recently to assist in the invalidation of the Hitler Diaries. Francois Demelle in France reportedly published the 1st study on questioned document examination in 1609, but it was not until the 1800s that document examination truly began in Europe. Photographers at that time attempted to broaden their professional expertise by providing services that incorporated document comparison, but highly publicized mistakes delayed the acceptance of the profession.

One such mistake involved Alphonse Bertillon, previously acclaimed for his invention and application of anthropometry to personal identification. In the Dreyfus case, Bertillon testified that the accused had written a document that was the basis for the charge of treason. The accused's innocence was later proven. Old English laws that stated that writings were inadmissible as standards for comparison unless the writings were in evidence in a prior case, also allegedly hindered the lack of initial acceptance of the questioned document profession in the United States. The acceptance of questioned document testimony in courts eventually became a reality in 1913.

Nowadays, the Questioned Documents Unit of the FBI crime laboratory examines all aspects of documentary evidence, including hand printing, typewriter ribbons, printers, watermarks, erasures, alterations, obliterations, safety paper, charred paper, graphic arts, plastic bags and product tampering. The unit maintains the database files, such as the Anonymous Letter File, Bank Robbery Note File, National Fraudulent Check File, Office Equipment File and Watermark File.

Soil examination

The main contributor to the development of forensic soil and particulate examination was, again, Edmond Locard. In 1929 he made the observation that it was almost impossible for anyone to participate in any activity without removing soil or dust particles from the site, as a result of contamination of the person's body, clothes, tools or vehicle. However, Locard credited Sherlock Holmes with the idea that botanical or earth materials could have important evidential value. Arthur Conan Doyle topics, published between 1893 and 1897, suggest that soil collected on a person at a crime scene could be used as evidence to place that person at that location. The first recorded use of this type of evidence was in Germany in 1904, by Dr. George Popp who developed and presented what is thought to be the 1st example of evidence in a criminal case using botanical materials. Throughout the 1920s Popp continued to pioneer the use of botanical identification in forensic work.

IDENTIFICATION AND ASSOCIATION OF HUMAN HAIR

Rudolph Virchow, a professor and prosecutor in Berlin, Germany, who established that a questioned hair from a particular crime suspect was indistinguishable from that of the victim, reported the 1st forensic investigation of human hair in 1861. By the early 1900s the significance of hair evidence in criminal investigations was routinely criticized and scrutinized by a number of medicolegal experts.

In 1906, Hugo Marx, an official in the State Medical and Prison Medical Examiner's office in Berlin, wrote a dissertation on the value of forensic hair examination and its role in personal identification. A text topic published at the turn of the century, from the founder of the Vienna school of forensic medicine, also contained a chapter on the investigation of hair.

In 1910 Victor Balthazard and his associate, Marcelle Lambert, produced he poil de I'homme et des animaux (The Hair of Man and Animals). In this work they describe techniques closely resembling those still used in hair comparison today – mainly microscopical analysis to compare and observe morphological features. In 1931, John Glaister published a study of mammalian hairs and wool and their relevance to criminal investigation.

The work contains approximately 1700 photomicrographs that show the structure of hair of humans and other mammals. Paul Kirk, criminalistics professor at the University of California at Berkeley, further developed hair analyses by improving the techniques of hair comparison, utilizing various physical and chemical properties in an attempt to facilitate the systematic individualization of human hair.

Despite, cutting-edge advances in various other specialties of forensic science, the methods, philosophy and significance of forensic hair examination have not changed greatly since the early 19th century. The Trace Evidence Unit of the FBI today holds extensive reference collections of human and animal hairs, natural and manmade fibers, feathers, ropes, cordage, woods and seeds.

Deoxyribonucleic acid profiling

Evolving from classical serology, DNA profiling could be considered the modern-day technique revolutionizing personal identification in forensic science. In the mid-1980s Sir Alex Jeffreys developed the techniques allowing the profile analysis of DNA. After publishing his achievements in Nature in 1985, Jeffreys was subsequently called upon to apply his techniques to solve the 1st crime in 1986. In combination with the British Home Office Forensic Science Service, his DNA profiling techniques were used to identify Colin Pitchfork as the murderer of Dawn Ashworth and Lynda Mann in Leicestershire, England.

Cetus Corporation furthered the developments of DNA profiling and molecular biology techniques in personal identification during the rest of the

1980s with the development of the poly-merase chain reaction. In 1987, not only was DNA profiling introduced for the 1st time to US criminal courts, but the admissibility of DNA evidence was also challenged, resulting in the development and implementation of appropriate accreditation, standardization and quality controls for DNA and forensic laboratories. During 1996, the FBI DNA Analysis Unit began using mitochondrial DNA analysis.

This type of analysis can be applied to small or degraded quantities of DNA from hair, bones, teeth and body fluids, allowing the examination of evidence that may not have been suitable for comparison prior to the development of this technique.

Accreditation

Finally, over the last 30 years or so, several organizations have emerged that have developed guidelines and regulations assuring forensic laboratory standardization, accreditation and certification as a guide to legal and forensic communities. Of the numerous organizations that exist, 3 have made significant impacts on the professional appearance of the field of forensic science.

The American Society for Testing and Materials (ASTM) committee E-30 on Forensic Sciences was created in 1970, with the purpose of standardizing methods and terminology unique to the field. To keep pace with the rapid technological advances within the field, standards are continuously updated.

Established in the mid 1970s, the American Society of Crime Laboratory Directors (ASCLD) is an international organization composed of crime laboratory directors whose mission is to promote (excellence through leadership in forensic science management). The ASCLD laboratory accreditation board (ASCLD/LAB) is a related organization that has published minimum standards for laboratories and their personnel to achieve, allowing formal accreditation by the organization.

Around the same time the American Board of Criminalistics developed a series of examinations allowing the certification of individual forensic scientists in their particular area of expertise. Although individual certification processes are currently voluntary, it is probable that courts will come to expect such quality assurances as the number of certified laboratories and individuals using standard techniques continues to grow.

DEATH: EXTINCTION OR CESSATION OF LIFE

Death is the extinction or cessation of life, but since life itself is difficult to define there is a reciprocal problem in defining death. The precise definition of death will always be a subject of controversy because it has social and religious aspects and is not a solely scientific issue. There are profound social and legal repercussions to the diagnosis. An added difficulty is that in nearly all circumstances human death is a process rather than an event. Within this process of dying there are points of no return, the identification of which is the medical diagnostic challenge.

DIAGNOSIS OF DEATH

There is no legal definition of death. The diagnosis of clinical death, or somatic death, is traditionally made using the triad of Bichat which states that death is «the failure of the body as an integrated system associated with the irreversible loss of circulation, respiration and innervation». Thus the diagnosis of death is made by excluding possible signs of life. The irreversible cessation of the circulation has been considered for centuries a point of no return. It still provides a practical and valid criterion for the irreversible loss of function of the human organism as a whole.

To ensure that opportunities for resuscitation are not missed, care must be taken in making this final diagnosis in order to avoid mistaking apparent death for actual death. In the overwhelming majority of deaths the diagnosis can be made by traditional clinical methods. The same criteria are applicable whether the death was expected or unexpected. Firstly, there is a need to make a rapid assessment, based upon a history and clinical observations, as to whether resuscitation attempts should be initiated. Cessation of the circulation results in a deathly pallor (pallor mortis) particularly of the face and lips, and primary muscular flaccidity leads to drooping of the lower jaw and sometimes open staring eyes.

A complete physical examination should exclude the presence of a circulation or breathing. The absence of a pulse should be determined through the palpation of the carotid, radial and femoral arteries. The absence of heart and lung sounds should be determined by auscultation continually for one minute and repeated intermittently over not less than five minutes. Normal heart sounds may be indistinct in obese individuals or conditions such as pericardial effusion.At the same time observation should be made for respiration. Inspection of the eyes should disclose pupils which are non-reactive to a bright light. Indisputable signs of death develop later with the formation of livor mortis and rigor mortis.

Some situations, most notably hypothermia, produce death like states. Other conditions that can induce a death-like coma include drug overdose (particularly with barbiturates, alcohol, tricyclic antidepressants and anaesthetic agents), and metabolic states including myxoedema coma, uraemia, hypoglycaemia, hyperosmolar coma, and hepatic encephalopathy. Situations in which vigorous attempts at resuscitation may be successful include drowning, airways obstruction, electric shock, and a lightning strike.

Brain-stem death

The cessation of the circulation is only lethal if it lasts long enough to cause critical centres in the brain-stem to die. This is so because the brain-stem is irreplaceable in a way in which the pumping function of the heart is not. Viewing death in terms of brain-stem death rather than the cessation of the circulation is a modern view of the previously recognised facts of the dying process. This reappraisal has been precipitated by modern technology in medicine. Today it is possible to have a body whose brain is irreversibly dead but whose ventilation is maintained by a respirator, cardiac function by various drugs or pumping devices, feeding by the intravenous route and the elimination of waste products by dialysis.

The diagnosis of brain-stem death only arises in a hospital setting in which a patient has suffered irreversible damage to the brain but breathing is being maintained by a ventilator and the heart is continuing to function. The brain-stem is that part of the brain at its base which includes the mesencephalon or midbrain, the pons and the medulla. Within the brain-stem are the respiratory and vasomotor centres which are responsible respectively for breathing and the maintenance of blood pressure. It also contains the ascending reticular activating system, which maintains alertness or the capacity for consciousness.

This capacity for consciousness, which is a function of the upper brain-stem, is not the same as the content of consciousness, which is a function of the cerebral hemispheres, but rather it is an essential pre-condition for the latter. Without the function of the brain-stem there can be no cognitive or affective life and no social interaction with the environment. The capacity to breathe is a brain-stem function and cessation of breathing, apnoea, is a critical manifestation of a non-functioning lower brain-stem. Once spontaneous breathing ceases the heart cannot continue to function for long and the circulation then ceases. Following judicial hangings which fracture the neck some cardiac activity can be maintained for up to a maximum of 20 minutes.

In 1967 the 1st heart transplantations were performed in which the hearts were harvested from beating-heart brain-dead donors. At that time there were no guidelines for the diagnosis of death in these heart donors. The following year Harvard medical school published criteria for the recognition of the «brain-death syndrome», and in 1981 a model statute, called the Uniform Determination of Death Act, was published in the United States. This defined death as either irreversible cessation of circulatory and respiratory functions or irreversible cessation of all functions of the entire brain.

This concept of «whole brain death», if taken literally, would mean that the detection of any activity by any means in any part of the brain would preclude a diagnosis of death. This difficulty was avoided in the United Kingdom when in 1976 brain death was defined as the complete and irreversible loss of function of the brain-stem. This definition is simple, reliable and robust. Internationally, medical opinion and practice has moved in a similar direction in accepting the concept of brain- stem death. In hospital practice the diagnosis of brain-stem death is not technically difficult and can be made on purely clinical grounds. The purpose of the examination is to establish irreversible loss of brain-stem function. Mindful of the need to protect the right to life, most countries insist that the diagnosis is made by senior physicians in appropriate specialities such as neurology, anaesthetics or intensive care medicine, who are not in any way associated with the potential use of the patient's organs for transplantation. The diagnosis of brain-stem death is made in three phases. The 1st step is to ascertain the cause of the coma and to establish that there is irremediable structural brain damage.

The brain damage is judged irremediable based upon its context, the passage of time and the failure of all attempts to remedy it. The 2nd step is to exclude all possible causes of reversible brain-stem dys-function, such as hypothermia, drug intoxication or severe metabolic disturbances. The 3rd and final step is to demonstrate the absence of all brain-stem reflexes and the fact that the patient cannot breathe however strong the stimulus. The first two steps may take up to 48 hours but the 3rd step, the testing of brain-stem function, takes less than half an hour.

The normal brain-stem reflex responses tested for are:

- Constriction of the pupils in response to light;
- Blinking in response to stimulation of the cornea;
- Grimacing in response to firm pressure applied to just above the eye-socket;
- Movement of the eyes in response to the ears being flushed with ice water and
- Coughing or gagging in response to a suction catheter being passed down the airway.

The ability of the patient to breathe is assessed by ensuring full oxygenation through breathing 100 % oxygen for several minutes and then disconnecting the ventilator while maintaining diffusion oxygenation into the trachea via a catheter. This test allows the carbon dioxide concentration in the blood to rise to levels more than sufficient to stimulate inspiratory effort. At the same time the patient is protected against serious oxygen deprivation while disconnected from the ventilator. Both apnoea and the absence of brain-stem reflexes must be confirmed twice.

Organ Transplantation

The ability to transplant organs from one human being to another is one of the great achievements of modern medicine. The 1st transplant was of a kidney from a living donor into her identical twin sister in 1954. The 1st liver transplant was in 1963, pancreas — in 1966, and heart — in 1967. In the 1960's, many organs were harvested from non-heartbeating corpses whose deaths had been certified on classical cardiopulmonary criteria. However, transplant of organs from these cadavers was less successful than using organs from live donors. This was because, with non-heartbeating cadavers, ischaemic injury to the organs began with the cessation of the circulation, so the organs suffered a relatively long ischaemic time before transplantation. As a consequence living-related and living-unrelated organ donation became common but gave rise to ethical concerns about the possibility of coercion of the donors and the potential for the sale of organs. With the development of the concept of brainstem death these heartbeating cadavers became the most common source of donor organs from the early 1970's onwards. Such organs have the benefit of being clinically equivalent to organs obtained from living donors, having equivalent ischaemic times, but do not bring the same ethical concerns of possible coercion or commercialisation.

Nowadays the problem which has arisen is that demand for organs has far outstripped the supply and a quarter of all patients awaiting organ donation may die before receiving a transplant. Maximising the donation rate from all potential brainstem dead donors would still not meet the demand for organs. As a result there is now a return to the use of organ donation from individuals pronounced dead using classical cardiopulmonary criteria, and of whom there is a potentially large pool, such as those dead on arrival at hospital or where in-hospital resuscitation is unsuccessful. There has been also a resurgence of living-related and living-unrelated donations of kidneys, liver and lung lobe for transplantation.

The forensic interest in organ transplantation arises because of the use of organs from donors whose death requires a medico-legal investigation, such as victims of road traffic accidents, industrial accidents, assaults and sudden unexpected deaths. In such cases the consent of the investigating authorities is necessary prior to organ donation. That consent is usually given since it is self-evident that any organs suitable for transplantation must be functioning normally, and that any subsequent medico-legal autopsy on the donor will not be compromised by an inability to dissect and examine the donated organs.

Permanent vegetative state

A permanent vegetative state is a clinical condition of unawareness of self and environment in which the patient breathes spontaneously, has a stable circulation, and shows cycles of eye closure and opening which may simulate sleep and wakening. A vegetative state can be regarded as permanent, 3 months after non-traumatic brain damage or 12 months after traumatic injury. In this condition the brain stem is mostly spared, whereas the grey and white matter of both cerebral hemispheres is widely and severely damaged.

Consciousness has 2 main components, namely arousal or wakefulness, and awareness of the environment and of the self. Arousal is a brain stem function while awareness is dependent upon the functional integrity of the cerebral cortex and its sub cortical connections. The vegetative state is a state of wakeful unawareness. Consequently the diagnosis of permanent vegetative state depends on providing evidence of a negative, a lack of awareness. The diagnosis is made when the patient shows no behavioural evidence of awareness of self or environment; there is brain damage, usually of a known cause consistent with the diagnosis; there are no reversible causes present; and at least 6 months, and usually 12 months, have passed since onset of the condition.

The 3 major sensory systems (auditory, visual and somatic) and the motor system are assessed to establish that some sensory stimuli can enter the central nervous system and that the motor pathway out is functioning but that there is no evidence of: (1) any spontaneous meaningful motor activity, including voice; (2) language comprehension or expression and (3) sustained, reproducible, purposeful, or voluntary behavioural responses to normal or noxious visual, auditory, or tactile stimuli.

Patients with the diagnosis of permanent vegetative state breathe spontaneously, because they have preserved brain stem function, and may live for many years if artificial nutrition and hydration is maintained. Stopping food and water inevitably leads to death within 14 days from dehydration, but doing so typically requires the consent of the court. The leading English case is that of Tony Bland, a young man diagnosed with permanent vegetative state as a result of crush asphyxiation in the Hillsborough (Liverpool) football stadium disaster of 1991. In 1993 the High Court approved withdrawal of artificial feeding and hydration and he died a week or so later. Since that landmark case there have been approximately 20 similar cases in England.

Patients in the permanent vegetative state raise ethical issues concerning the nature of consciousness, quality of life and the value society attributes to life. The diagnosis is difficult because there is no definitive test for awareness and the biology of consciousness is not understood. This is in marked contrast with the concept of brain stem death where both the anatomy and the physiology are well understood.

Cellular death

Clinical death represents somatic death, that is to say the death of a person as a whole. However, not all the cells of the body die at the same time. For some hours after death the pupils will still respond to pilocarpine drops by contracting, and electrical stimulation of muscles will cause contraction. The cornea of the eye may still be suitable for transplant up to 24 hours after death. Viable skin grafts can be obtained for up to 24 hours, bone grafts for up to 48 hours and arterial grafts for up to 72 hours after the circulation has stopped.

Post mortem the cells of the body are destroyed by the process of autolysis (literally «self-destruction»), with waves of cell death following somatic death. Destructive enzymes released from lysosomes within the cell initiate the process of autolysis. The process is more rapid in some organs, for example in the pancreas which also contains a large number of digestive enzymes normally secreted into the gut. At a microscopic level autolysis is evidenced by a homogenous staining of the cytoplasm of the cell and similar loss of characteristic staining and detail within the nucleus.

This post mortem change occurring in all the cells of the body is similar to the change which occurs in damaged cells in a living body. Within a living person individual cells or large areas of tissue, comprising groups of adjacent cells, may die without affecting the viability of the whole organism. This pathological cell death, or necrosis, is an abnormal change initiated by some insult to the tissues, such as hypoxia or physical or chemical trauma. Within a few hours of being irreversibly damaged the cells show the microscopic changes characteristic of autolysis.

However, unlike post mortem autolysis, this type of cell death, necrosis, incites an inflammatory reaction from the surrounding living tissue. It is the presence of this inflammatory reaction, which can be identified microscopically, which distinguishes tissue necrosis which has occurred in life from post mortem autolysis. Unfortunately the inflammatory reaction only develops to a level at which it can be identified microscopically between A hours and 2 hours after injury. Consequently injuries which are inflicted very shortly before death, like tissue damage inflicted after death, show no vital inflammatory reaction. However, the degree of bruising of the tissues associated with the injury may give an indication as to whether or not there was a functioning circulation.

In the living body the cells of all tissues turnover with loss of some cells and their replacement by more cells created by mitotic division. Apoptosis (meaning dropping out or falling away) describes the energy dependant process by which individual cells are lost. The cell contracts and the nucleus fragments producing an apoptotic body, pieces of which are removed by scavenger cells, macrophages. Apoptosis is a normal process and does not stimulate an inflammatory reaction from the adjacent tissues. It is important in the natural turnover of many tissues such as the endometrium during the menstrual cycle. Unlike necrosis and autolysis, it has no forensic importance.

PRINCIPAL POST MORTEM CHANGES

The 3 principal post mortem changes which occur within the 1st day after death are: body cooling (algor mortis, literally: «the chill of death»), livor mortis (literally: «the darkening of death»), and rigor mortis (literally: «the stiffening of death»). Putrefaction of the body and its variants are later post mortem changes. These changes which develop in a corpse well after death has occurred are of interest for several reasons. They are indisputable signs of death and indicate that any attempts at resuscitation would be futile. As they evolve these post mortem changes produce confusing artefacts and putrefaction destroys evidence of identity, injuries, and natural disease. However, each has its own specific forensic uses. Since they all evolve over time they all have been used also to estimate the time since death. The importance of body cooling lies solely in its value for the estimation of the time since death, and therefore it is discussed in that context in the next chapter.

RIGOR MORTIS

Death is followed immediately by total muscular relaxation, primary muscular flaccidity, so that the body collapses into a position dictated by gravity and surrounding objects. Flaccidity is succeeded in turn by generalised muscular stiffening, rigor mortis, which fixes the body in that posture. It follows that rigor cannot freeze a body in a position which defies gravity, and any such appearance indicates that the body has been moved after rigor developed. If the body is supine then the large joints of the limbs become slightly flexed during the development of rigor. The joints of the fingers and toes are often markedly flexed due to the shortening of the muscles of the forearms and legs.

After a variable period of time, as a result of putrefaction, rigor mortis passes off to be followed by secondary muscular flaccidity. There is great variation in the rate of onset and the duration of rigor mortis; the two main influencing factors are the environmental temperature and the degree of muscular activity before death. Onset of rigor is accelerated and its duration shortened when the environmental temperature is high and after prolonged muscular activity, e.g. following convulsions. Conversely, a late onset of rigor in many sudden deaths can be explained by the lack of muscular activity immediately prior to death.

Classically, rigor is said to develop sequentially, but this is not constant or symmetrical. Typically rigor is apparent first in the small muscles of the eyelids, lower jaw and neck, followed by the limbs. It involves first the small distal joints of the hands and feet, and then the larger proximal joints of the elbows and knees, and then the shoulders and hips. Ante-mortem exertion usually causes rigor to develop first in the muscles used in the activity.

Generally rigor passes off in the same order in which it develops. Gently attempted flexion of the different joints will indicate the location of rigor and its degree (complete, partial, or absent joint fixation), providing no artefact has been introduced by previous manipulation of the body by others, such as during the removal of clothing. The forcible bending of a joint against the fixation of rigor results in tearing of the muscles and the rigor is said to have been «broken». Provided the rigor had been fully established, it will not reappear once broken down by force. The intensity or strength of rigor mortis depends upon the decedent's muscular development, and should not be confused with its degree of development, that is the extent of joint fixation.

Rigor involves voluntary and involuntary muscles. Rigor of the myocardium should not be mistaken for myocardial hypertrophy. Likewise secondary muscular flaccidity of the ventricles should not be mistaken for ante-mortem dilatation or evidence of myocardial dysfunction. Involvement of the iris muscles means that the state of the pupils after death is not a reliable indication of their ante-mortem appearance. Different degrees of rigor can cause irregularity and inequality of the pupils. Contraction of the arrectores pilorum muscles during rigor causes «goose-flesh» (cutis anserina), a phenomenon commonly seen in bodies recovered from water. Involvement of the walls of the seminal vesicles by rigor may lead to discharge of seminal fluid at the glans penis.

The biochemical basis of rigor mortis is not fully understood. Post-mortem loss of integrity of the muscle cell sarcoplasmic reticulum allows calcium ions to flood the contractile units (sarcomeres) initiating the binding of actin and myosin molecules and mimicking the normal contraction process. Normal relaxation in life is achieved by energy-dependent (ATP-driven) pumping of calcium back across the membrane of the sarcoplasmic reticulum but this fails post-mortem because of membrane disruption and lack of ATP.

The actin-myosin complex is trapped in a state of contraction until it is physically disrupted by the autolysis which heralds the onset of putrefaction. This process is characterised by proteolytic detachment of actin molecules from the ends of the sarcomeres, and consequent loss of the structural integrity of the contractile units. Although the biochemical basis of rigor mimics that of muscle contraction in life, it does not cause any significant movement of the body in death, a point of forensic importance.

Cadaveric spasm

Cadaveric spasm (synonyms: instantaneous rigor, instantaneous rigidity, cataleptic rigidity) is a form of muscular stiffening which occurs at the moment of death and which persists into the period of rigor mortis. Its cause is unknown but it is usually associated with violent deaths in circumstances of intense emotion. It has medico-legal significance because it records the last act of life. Cadaveric spasm involving all the muscles of the body is exceedingly rare and most often described in battle situations.

Most commonly cadaveric spasm involves groups of muscles only, such as the muscles of the forearms and hands. Should an object be held in the hand of a corpse, then cadaveric spasm should only be diagnosed if the object is firmly held and considerable force is required to break the grip. This is seen in a small proportion of suicidal deaths from firearms, incised wounds, and stab wounds, when the weapon is firmly grasped in the hand at the moment of death.

In such circumstances the gripping of the weapon creates a presumption of self-infliction of the injuries. This state cannot be reproduced after death by placing a weapon in the hands. It is also seen in cases of drowning when grass, weeds, or other materials are clutched by the deceased. Similarly, in mountain falls, branches of shrubs or trees may be seized. In some homicides, hair or clothing of the assailant can be found gripped in the hands of the deceased.

Livor mortis

Lividity is a dark purple discoloration of the skin resulting from the gravitational pooling of blood in the veins and capillary beds of the dependent parts of the corpse. Synonyms include: livor mortis, hypostasis, post-mortem lividity, and, in the older literature, post-mortem suggillations. Lividity is able to develop post mortem under the influence of gravity because the blood remains liquid rather than coagulating throughout the vascular system as a consequence of stasis. Within about 30 to 60 minutes of death the blood in most corpses becomes permanently incoagulable. This is due to the release of fibrinolysins, especially from capillaries and from serous surfaces, e. g. the pleura. The fluidity and incoagulability of the blood is a commonplace observation at autopsy and is not characteristic of any special cause or mechanism of death.

Hypostasis begins to form immediately after death, but it may not be visible for some time. Ordinarily its earliest appearance, as dull red patches, is 20 to 30 minutes after death, but this may be delayed for some hours. Faint lividity may appear shortly before death in individuals with terminal circulatory failure. Conversely, the development of lividity may be delayed in persons with chronic anaemia or massive terminal haemorrhage. The patches of livor then increase in intensity and become confluent to reach a maximum extent and intensity on average within about 12 hours, although there is very great variation.

Pressure of even a mild degree prevents the formation of lividity in that area of skin, so that a supine body shows contact flattening associated with contact pallor (pressure pallor) over the shoulder blades, elbows, buttocks, thighs and calves. Similarly tight areas of clothing or jewellery, as well as skin folds, leave marks of contact pallor. The distribution of lividity with its associated contact pallor helps distinguish lividity from bruising, and any doubts are resolved by incising the skin which reveals lividity as congested vessels and bruising as haemorrhage infiltrating tissues.

Lividity is present in all corpses, although it may be inconspicuous in some, such as following death from exsanguination. Lividity is usually well marked in the earlobes and in the fingernail beds. In a supine corpse there may be isolated areas of lividity over the front and sides of the neck resulting from incomplete emptying of superficial veins. Other isolated patches of hypostasis may be due to blood in the deeper veins being squeezed, against gravity, towards the skin surface by the action of muscles developing rigor mortis.

Lividity is often associated with post-mortem haemorrhagic spots, punctate haemorrhages (given the specific name «vibices» in the German literature), which resemble the petechial haemorrhages associated with asphyxial deaths, and from which they must be distinguished. Easily recognized, occurring only in areas of lividity and sparing adjacent areas of contact pallor, they develop in the hours immediately following death as lividity intensifies.

Lividity occurs in the viscera as well as the skin and this provides some confirmation of the external observations. In the myocardium lividity may be mistaken for an acute myocardial infarction, and in the lungs may be misdiagnosed as pneumonia. Livid coils of intestine may falsely suggest haemorrhagic infarction. Lividity developing in the viscera of a body lying prone and resulting in a purplish congestion of organs usually found pale at autopsy can be disconcerting to those unaccustomed to these changes.

The importance of lividity lies in its distribution, as an indicator of body position and contact with objects, and in its colour, as an indicator of cause of death. The usual purple colour of lividity reflects the presence of deoxyhaemoglobin but it does not have the same diagnostic significance as cyanosis produced during life. In the corpse, oxygen dissociation from oxyhaemoglobin continues after death and there may be reflux of deoxygenated venous blood into the capillaries.

For these reasons, the blood of a cadaver becomes purplish-blue, but this is not a reflection of a pathophysiological change which occurred in life. Bodies refrigerated very soon after death have a pink lividity due to retained oxyhaemoglobin. Death from hypothermia or cyanide poisoning also imparts the pink hue of oxyhaemoglobin, carbon-monoxide poisoning produces the cherry red of carboxyhaemoglobin, and poisoning from sodium chlorate, nitrates and aniline derivatives impart the gray to brown colour of methaemoglobin. Infection by Clostridium perfringens causing gas gangrene is said to give a bronze lividity.

After about 12 hours lividity becomes «fixed» and repositioning the body, e. g. from the prone to the supine position, will result in a dual pattern of lividity since the primary distribution will not fade completely but a secondary distribution will develop in the newly dependent parts. The blanching of livor by thumb pressure is a simple indicator that lividity is not fixed. Fixation of lividity is a relative, not an absolute, phenomenon. Well-developed lividity fades very slowly and only incompletely. Fading of the primary pattern and development of a secondary pattern of lividity will be quicker and more complete if the body is moved early during the 1st day.

However, even after a post-mortem interval of 24 hours, moving the body may result in a secondary pattern of lividity developing. Duality of the distribution of lividity is important because it shows that the body has been moved after death. However, it is not possible to estimate with any precision, from the dual pattern of livor, when it was that the corpse was moved. If a prone body is moved some hours after death but before lividity is fixed then the primary lividity will fade and may leave behind on the face any lividity-associated punctuate haemorrhages, or «vibices», creating possible confusion with the petechiae of asphyxia.

Areas of lividity are overtaken early in the putrefactive process becoming green at first and later black. The red cells are haemolysed and the haemoglobin stains the intima of large blood vessels and diffuses into the surrounding tissues, highlighting the superficial veins of the skin as a purple-brown network of arborescent markings, an appearance referred to as «marbling».

Putrefaction

Putrefaction is the post-mortem destruction of the soft tissues of the body by the action of bacteria and endogenous enzymes and is entirely capable of skeletonising a body. Refrigeration of a corpse delays the onset of putrefaction, freezing the body halts putrefaction, and chemical embalming prevents it. The main changes recognizable in tissues undergoing putrefaction are the evolution of gases, changes in colour and liquefaction. These same changes seen on the surface of the body occur simultaneously in the internal organs. Bacteria are essential to putrefaction and commensal bacteria, mainly from the large bowel, soon invade the tissues after death.

Typically, the 1st visible sign of putrefaction is a greenish discoloration of the skin of the anterior abdominal wall due to sulph-haemoglobin formation. This most commonly begins in the right iliac fossa, i. e. over the area of the caecum. Any ante-mortem bacterial infection of the body, particularly scepticaemia, will hasten putrefaction. Injuries to the body surface promote putrefaction by providing portals of entry for bacteria. Putrefaction is delayed in deaths from exsanguination because it is blood which usually provides a channel for the spread of putrefactive organisms within the body.

Environmental temperature has a very great influence on the rate of development of putrefaction, so that rapid cooling of the body following a sudden death will markedly delay its onset. In a temperate climate the degree of putrefaction reached after 24 hours in the height of summer may require 10 to 14 days in the depth of winter. Putrefaction is optimal at temperatures ranging between 21 and 38°C (70 and 100°F), and is retarded when the temperature falls below 10°C (50°F) or when it exceeds 38°C (100°F). Heavy clothing and other coverings, by retaining body heat, will speed up putrefaction. The rate of putrefaction is influenced by body build because this affects body cooling. Obese individuals putrefy more rapidly than those who are lean.

Gases produced by putrefaction include methane, hydrogen, hydrogen sulphide and carbon dioxide. The sulphur-containing amino acids, cysteine, cystine and methionine yield hydrogen sulphide, which combines with haemoglobin and ferrous iron to produce green sulph-haemoglobin and black ferrous sulphide respectively. De-carboxylation of the amino acids ornithine and lysine yields carbon dioxide and the foul smelling ptomaines, putrescine (1,4-butanediamine) and cadaverine (1,5-pentanediamine) respectively. These ptomaines are detectable by the cadaver dogs used to locate clandestine graves. Deamination of L-phenylanaline yields ammonia, and phenylpyruvic acid which forms a green complex with ferric iron. Bacterial and fungal fermentation yield ethyl alcohol (ethanol), confounding the interpretation of post-mortem alcohol concentrations.

Early putrefaction is heralded by the waning of rigor, green abdominal discolouration, a doughy consistency to the tissues and haemolytic staining of vessels. Localised drying of the lips, tip of the nose and fingers may be seen. The face swells and discolours and the swollen lips are everted, making facial recognition unreliable. The skin, which now has a glistening, dusky, reddish-green to purple-black appearance, displays slippage of large sheets of epidermis after any light contact with the body, e.g. during its removal from the scene of death. Beneath the shed epidermis is a shiny, moist, pink base which dries, if environmental conditions permit, to give a yellow parchmented appearance. This putrefactive skin-slip superficially resembles ante-mortem abrasions and scalds. Body hair and nails are loosened and the skin of the hands comes away like gloves taking with it fingerprint evidence of identity. The remaining dermis has a much shallower reverse print which is technically more difficult to document.

Distention of the abdominal cavity by putrefactive gasses characterises the bloating stage of decomposition. In males gas is forced from the peritoneal space down the inguinal canals and into the scrotum, resulting in massive scrotal swelling. Gaseous pressure expels dark malodorous fluid, purge fluid, from the nose and mouth, mimicking ante-mortem haemorrhage or injury. Similar fluid flows from the vagina and anus, the rectum is emptied of faeces and prolapse of the rectum and uterus may occur.

The doughy consistency of the tissues of early putrefaction is replaced by the crepitant effect resulting from gaseous infiltration beneath the skin and in deeper tissues. Large sub-epidermal bullae fill with gas, sanguinous fluid or clear fluid. Gas bubbles appear within solid organs such as liver and brain giving a «Swiss-cheese» appearance, and the blood vessels and heart are filled with gas. These putrefactive changes are relatively rapid when contrasted with the terminal decay of the body. The more dense fibro-muscular organs such as the prostate and uterus remain recognizable until late in the process, thus aiding in the identification of sex. When the putrefactive juices have drained away and the soft tissues have shrunk, the speed of decay is appreciably reduced.

The progression of putrefaction may be modified by vertebrate or invertebrate animal activity. Wild animals, domestic pets, livestock, fish and crustaceans may be involved but most commonly it is insects, particularly fly larvae (maggots). In a hot humid environment with heavy insect activity a corpse may be skeletonised in as little as 3 days. All soft tissues are generally lost before the skeleton becomes disarticulated, typically from the head downward, with the mandible separating from the skull and the head from the vertebral column, and from central to peripheral, i. e. from vertebral column to limbs.

Mummification

Mummification is a modification of putrefaction characterised by the dehydration or dessication of the tissues. The body shrivels and is converted into a leathery or parchment-like mass of skin and tendons surrounding the bone. Skin shrinkage may produce large artefactual splits mimicking injuries, particularly in the groins, neck, and armpits. Mummification develops in conditions of dry heat, especially when there are air currents, e. g. in a desert. Mummification of bodies in temperate climates is unusual unless associated with forced hot-air heating in buildings or other man-made favourable conditions. The importance of mummification lays in its preservation of tissues which aids in personal identification and the recognition of injuries. However, mummified tissues may be attacked by rodents and insects, particularly the omnivorous larvae of the brown house moth (Hofmannophila pseudospretella) which is found in many countries worldwide.

Adipocere

Adipocere formation, or saponification, is a modification of putrefaction characterised by the transformation of fatty tissues into a yellowish-white, greasy, wax-like substance which is friable when dry. During the early stages of its production it has a very persistent ammoniacal smell but once its formation is complete it has a sweetish rancid odour. Adipocere, also known as «grave wax» or «corpse wax», develops as the result of hydrolysis of fat with the release of fatty acids which, being acidic, inhibit putrefactive bacteria. Fatty acids combine with sodium or potassium to form hard soap or soft soap («sapo domesticus») respectively. Calcium gives an insoluble soap which contributes a more brittle quality to the adipocere.

However, fat and water alone do not produce adipocere. Putrefactive organisms, of which Clostridium welchii is most active, are important, and adipocere formation is facilitated by post-mortem invasion of the tissues by commensal bacteria. A warm, moist, anaerobic environment favours adipocere formation. Adipocere develops first in the subcutaneous tissues, most commonly involving the cheeks, breasts and buttocks. Rarely, it may involve the viscera such as the liver. The adipocere is admixed with the mummified remains of muscles, fibrous tissues and nerves. Putrefaction, adipocere and mummification may coexist in the same corpse or in adjacent corpses within mass graves as a consequence of differing micro-environments. The importance of adipocere lies in its preservation of the body, which aids in personal identification and the recognition of injuries.

Maceration

Maceration is the aseptic autolysis of a foetus, which has died in-utero and remained enclosed within the amniotic sac. Bacterial putrefaction plays no role in the process. The changes of maceration are only seen when a still-born foetus has been dead for several days before delivery. Examination of the body needs to be prompt since bacterial putrefaction will begin following delivery. The body is extremely flaccid with a flattened head and undue mobility of the skull. The limbs may be readily separated from the body. There are large moist skin bullae, which rupture to disclose a reddish-brown surface denuded of epidermis. Skin slip discloses similar underlying discoloration. The body has a rancid odour but there is no gas formation. Establishing maceration of the foetus provides proof of a post-mortem interval in-utero, and therefore proof of stillbirth and conclusive evidence against infanticide.

AN ESTIMATE OF THE TIME OF DEATH

All death certificates require an estimate of the time of death as well as a statement of the underlying cause of death. If the death was witnessed then providing a time of death presents no difficulty, but in an un-witnessed death it can be problematic. Establishing the time of death is of assistance in any police investigation of a death, whether it was from natural or un-natural causes. Establishing the time of an assault and the time of death is critical in criminal proceedings in which there are legal issues of alibi and opportunity to commit the crime. If an accused can prove that he was at some other place when the injury and death of the victim occurred then his innocence is implicit. However, the time of injury, or indeed the time of onset of an acute illness, may be separated from the time of death by a significant survival period.

Evidence of the time elapsed since death, the post-mortem interval, may come from the body of the deceased, from the environment in the vicinity of the body, and from information on the deceased's habits, movements, and day-to-day activities. All 3 sources of evidence — corporal, environmental and anamnestic — should be explored and assessed before offering an opinion on when death occurred. The longer the post-mortem interval then the less accurate is the estimate of it based upon corporal changes. As a consequence, the longer the post-mortem interval then the more likely it is that anamnestic or environmental evidence will provide the most reliable estimates of the time elapsed.

Many physico-chemical changes begin to take place in the body immediately or shortly after death and progress in a fairly orderly fashion until the body disintegrates. Each change progresses at its own rate which, unfortunately, is strongly influenced by largely unpredictable endogenous and environmental factors. Consequently, using the evolution of post-mortem changes to estimate the post-mortem interval is invariably difficult, and always of limited accuracy.

BODY COOLING

Body cooling is the most useful single indicator of the postmortem interval during the first 24 hours after death. The use of this method is only possible in cool and temperate climates, because in tropical regions there may be a minimal fall in body temperature post-mortem, and in some extreme climates, such as desert regions, the body temperature may even rise after death.

Since body heat production ceases soon after death but loss of heat continues, the body cools. The fall in body temperature after death is mainly the result of radiation and convection. Evaporation may be a significant factor if the body or clothing is wet, and heat loss by conduction may be considerable if the body is lying on a cold surface. Newton's law of cooling states that the rate of cooling of an object is determined by the difference between the temperature of the object and the temperature of its environment, so that a graphical plot of temperature against time gives an exponential curve.

However, Newton's law applies to small inorganic objects and does not accurately describe the cooling of a corpse which has a large mass, an irregular shape, and is composed of tissues of different physical properties. The cooling of a human body is best represented by a sigmoid curve when temperature is plotted against time. Thus, there is an initial maintenance of body temperature which may last for some hours — the so-called «temperature plateau» — followed by a relatively linear rate of cooling, which subsequently slows rapidly as the body approaches the environmental temperature. The post-mortem temperature plateau is physically determined and is not a special feature of the dead human body. Any inert body with a low thermal conductivity has such a plateau during its early cooling phase. The post-mortem temperature plateau generally lasts between a half and 1 hour, but may persist for as long as 3 hours, and some authorities claim that it may persist for as long as 5 hours.

It is usually assumed that the body temperature at the time of death was normal i. e. 37°C. However, in individual cases the body temperature at death may be subnormal or markedly raised. As well as in deaths from hypothermia, the body temperature at death may be sub-normal in cases of congestive cardiac failure, massive haemorrhage, and shock. The body temperature may be raised at the time of death following an intense struggle, in heat stroke, in some infections, and in cases of haemorrhagic stroke involving the pons. Where there is a fulminating infection, e. g. septicaemia, the body temperature may continue to rise for some hours after death.

Thus the 2 important unknowns in assessing time of death from body temperature are the actual body temperature at the time of death, and the actual length of the post-mortem temperature plateau. For this reason assessment of time of death from body temperature cannot be accurate in the 1st 4 to 5 hours after death when these 2 unknown factors have a dominant influence. Similarly, body temperature cannot be a useful guide to time of death when the cadaveric temperature approaches that of the environment. However, in the intervening period, over the linear part of the sigmoid cooling curve, any formula which involves an averaging of the temperature decline per hour may well give a reasonably reliable approximation of the time elapsed since death. It is in this limited way that the cadaveric temperature may assist in estimating the time of death in the early post mortem period.

Unfortunately the linear rate of post-mortem cooling is affected by environmental factors other than the environmental temperature and by cadaveric factors other than the body temperature at the time of death. The most important of these factors are body size, body clothing or coverings, air movement and humidity, and wetting or immersion in water. Body size is a factor because the greater the surface area of the body relative to its mass, the more rapid will be its cooling. Consequently, the heavier the physique and the greater the obesity of the body, the slower will be the heat loss.

Children lose heat more quickly because their surface area to mass ratio is much greater than for adults. The exposed surface area of the body radiating heat to the environment will vary with the body position. If the body is supine and extended, only 80 % of the total surface area effectively loses heat, and in the foetal position the proportion is only 60 %. Clothing and coverings insulate the body from the environment and therefore slow body cooling. The effect of clothing has a greater impact on corpses of low body weight. A bedspread covering may at least halve the rate of cooling. For practical purposes, only the clothing or covering of the lower trunk is relevant.

Air movement accelerates cooling by promoting convection, and even the slightest sustained air movement is significant if the body is naked, thinly clothed or wet. Cooling is more rapid in a humid rather than a dry atmosphere because moist air is a better conductor of heat. In addition the humidity of the atmosphere will affect cooling by evaporation where the body or its clothing is wet. A cadaver cools more rapidly in water than in air because water is a far better conductor of heat. For a given environmental temperature, cooling in still water is about twice as fast as in air, and in flowing water, about 3 times as fast.

Simple formulae for estimating the time of death from body temperature are now regarded as naive. The best tested and most sophisticated current method for estimating the postmortem interval from body temperature is that of the German researcher Henssge. Even so, it is acknowledged that the method may produce occasional anomalous results. It uses a nomogram based upon a complex formula, which approximates the sigmoid-shaped cooling curve. To make the estimate of post-mortem interval, using this method requires: (a) the body weight; (b) the average environmental temperature since death and (c) the core body temperature measured at a known time, and assumes a normal body temperature at death of 37,2°C.

Empiric corrective factors allow for the effect of important variables such as clothing, wetting and air movement. At its most accurate this sophisticated methodology provides an estimate of the time of death within a time span of 5,6 hours with 95 % probability. Gathering the data necessary to use this method for estimating time of death means that the body temperature should be recorded as early as conveniently possible at the scene of death. The prevailing environmental temperature should also be recorded at the same time, and a note made of the environmental conditions at the time the body was first discovered, and any subsequent variation in those conditions. Measuring the body core temperature requires a direct measurement of the intra-abdominal temperature.

Oral and axillary temperatures of a corpse do not reflect the core temperature and cannot be used. Either the temperature is measured rectally, or the intra-hepatic or sub-hepatic temperature is measured through an abdominal wall stab. An ordinary clinical thermometer is useless because its range is too small and the thermometer is too short. A chemical thermometer 10 to 12 inches (25 to 30 cm) long with a range from 0 to 50 °C is ideal. Alternatively a thermocouple probe may be used and this has the advantage of a digital readout or a printed record.

Whether the temperature is measured via an abdominal stab or per rectum is a matter of professional judgement in each case. If there is easy access to the rectum without the need to seriously disturb the position of the body and if there is no reason to suspect sexual assault, then the temperature can be measured per rectum. It may be necessary to make small slits in the clothing to gain access to the rectum, if the body is clothed and the garments cannot be pushed to one side. The chemical thermometer must be inserted about 4 inches (10 cm) into the rectum and read in situ. The alternative is to make an abdominal stab wound after displacing or slitting any overlying clothing. The stab is made over the right lower ribs and the thermometer inserted within the substance of the liver, or alternatively a right subcostal stab will allow insertion of the thermometer onto the undersurface of the liver.

These temperature readings from the body represent data, which if not collected at the scene of death is irretrievably lost. Therefore the decision not to take such readings is always a considered one. If sequential measurements of body temperature are taken then the thermometer should be left in situ during this time period. Taking sequential readings is much easier with a thermo-couple and an attached print-out device.

Supravital reactivity

The fact that cellular death occurs in waves within the body tissues following somatic death is evidenced not only by the possibility of organ and tissue transplantation, but also by the persisting excitability of muscle after death, supravital reactivity. Skeletal muscle may be induced to contract in a corpse using mechanical stimulation or electrical stimulation. Mechanical excitation of a variety of muscles in the limbs and face can be achieved by striking them in the immediate post mortem period but the times at which this excitability is lost is not sufficiently well documented to be of forensic use in the determination of the time of death. The testing of electrical excitability of skeletal muscle requires specific electrical apparatus and the insertion of needles into the muscle. Using this technique on the facial muscles some reaction may be obtained up to 22 hours after death.

In practice two tests for the mechanical excitability of skeletal muscle in a corpse are of forensic value. Striking the lower third of the quadriceps femoris muscle about 4 inches (10 cm) above the patella causes an upward movement of the patella because of a contraction of the whole muscle. If present this reaction indicates death within 2 hours. It is described as Xsako's phenomenon after the person who first described it. Similarly, striking the biceps brachii muscle and producing a muscular bulge at the point of impact, due to local contraction of the muscle, indicates that death had occurred within 13 hours. The absence of muscle contraction in either test provides no useful information.

In the post mortem period the smooth muscle of the iris is reactive to electrical and chemical stimulation for a longer time than skeletal muscle. The early death of the cells of the nervous system effectively denervates the smooth muscle of the iris, which becomes super-sensitive to chemicals which act at the neuromuscular junction. A change in the size of the pupil of the eye of a corpse can be produced by chemical stimulation of the iris following sub-conjunctival injection of solutions of acetylcholine, noradrenaline, and atropine. The strongest and longest surviving post mortem chemical stimulation is by acetylcholine and noradrenaline, the former producing miosis (papillary contraction) and the latter mydriasis (papillary enlargement). Reactivity to these two chemical neurotransmitters is lost at the earliest 14 hours after death and persists at the latest until 46 hours after death. Atropine produces mydriasis; the reactivity is lost by 3 hours post mortem at the earliest and is present until 10 hours post mortem at the latest.

Biochemical methods

A wide range of biochemical tests have been explored in an attempt to find one of use in estimating time of death, but without any success. This is not surprising since all postmortem biochemical changes will be temperature dependent and therefore less reliable than the use of body temperature itself in time of death estimation.

The biochemical method most frequently referred to is the measurement of potassium in the vitreous humour of the eye. There are sampling problems because the potassium concentration may differ significantly between the left and right eye at the same moment in time. The confounding effect of possible ante-mortem electrolyte disturbances can be excluded by eliminating all cases with a vitreous urea above an arbitrary level of 100 mg/dl, since high urea values in vitreous humour always reflect ante-mortem retention and are not due to post mortem changes. Having eliminated cases with possible ante-mortem electrolyte imbalance, there is a linear relationship between potassium concentration and time after death up to 120 hours, but the 95 % confidence limits are ± 22 hours, so the method is too imprecise to have practical value.

Rigor mortis

There is great variation in the rate of onset and the duration of rigor mortis, so that using the state of rigor mortis to estimate the post-mortem interval is of very little value. In general, if the body has cooled to the environmental temperature and rigor is well developed, then death occurred more than 1 day previously and less than the time anticipated for the onset of putrefaction, which is about 3 to 4 days in a temperate climate. Gently attempting flexion of the different joints will indicate the degree and location of rigor.

Typically slight rigor can be detected within a minimum of one half hour after death but may be delayed for up to 7 hours. The average time of first appearance is 3 hours. It reaches a maximum, i. e. complete development, after an average 8 hours, but sometimes as early as 2 hours post-mortem or as late as 20 hours. As a general rule when the onset of rigor is rapid, then its duration is relatively short. The 2 main factors which influence the onset and duration of rigor are the environmental temperature and the degree of muscular activity before death. Onset of rigor is accelerated and its duration shortened when the environmental temperature is high, so that putrefaction may completely displace rigor within 9 to 12 hours of death.

The forcible bending of a joint against the fixation of rigor results in tearing of the muscles and the rigor is said to have been «broken». Provided the rigor had been fully established, it will not reappear once broken down by force. Re-establishment of rigor, albeit of lesser degree, after breaking it suggests that death occurred less than about 8 hours before rigor was broken.

Livor Mortis

The development of livor is too variable to serve as a useful indicator of the post-mortem interval. Lividity begins to form immediately after death, but it may not be visible for some time. Ordinarily its earliest appearance, as dull red patches, is 20 to 30 minutes after death, but this may be delayed for up to 2, or rarely 3 hours. The patches of livor then deepen, increase in intensity, and become confluent within 1 to 4 hours postmortem, to reach a maximum extent and intensity within about 6 to 10 hours, but sometimes as early as 3 hours or as late as 16 hours. Faint lividity may appear shortly before death in individuals with terminal circulatory failure. Conversely, the development of lividity may be delayed in persons with chronic anaemia or massive terminal haemorrhage.

Putrefaction

There is considerable variation in the time of onset and the rate of progression of putrefaction. As a result, the time taken to reach a given state of putrefaction cannot be judged with accuracy. An observer should not assert too readily that the decomposed state of a body is inconsistent with a time interval alleged. As a general rule, when the onset of putrefaction is rapid then the progress is accelerated. Under average conditions in a temperate climate the earliest putrefactive changes involving the anterior abdominal wall occur between about 36 hours and 3 days after death. Progression to gas formation, and bloating of the body, occurs after about one week. The temperature of the body after death is the most important factor determining the rate of putrefaction. If it is maintained above 26°C (80°F) or so then the putrefactive changes become obvious within 24 hours and gas formation is seen in about 2 to 3 days.

The progression of putrefaction may be modified by vertebrate or invertebrate animal activity. Wild animals, domestic pets, livestock, fish and crustaceans may be involved but most commonly it is insects, particularly fly larvae (maggots). In a hot humid environment with heavy insect activity a corpse can be skeletonised in as little as 3 days. All soft tissues are generally lost before the skeleton begins to disarticulate, typically from the head downward, with the mandible separating from the skull and the head from the vertebral column, and from central to peripheral, that is from vertebral column to limbs. Remnants of ligaments and tendons commonly survive about one year, and an odour of decomposition for a few years.

Skeletal remains are of forensic interest only if the time since death is less than a human lifespan, about 75 years, because any perpetrator of a crime may still be alive. Dating skeletons is difficult but is aided by associated artefacts, such as personal effects, and evidence from the grave and its environment. Usefully the bones of individuals who died after the 1940s contain high levels of strontium-90 acquired in life from the atmospheric contamination caused by nuclear explosions.

The presence of any adipocere indicates that the post-mortem interval is at least weeks and probably several months. Under ideal warm, damp conditions, adipocere may be apparent to the naked eye after 3–4 weeks. Ordinarily, this requires some months and extensive adipocere is usually not seen before 5 or 6 months after death. Extensive changes may require not less than a year after submersion, or upwards of 3 years after burial. Once formed, adipocere will ordinarily remain unchanged for years.

Mummification develops in conditions of dry heat, especially when there are air currents. The time required for complete mummification of a body cannot be precisely stated, but in ideal conditions mummification may be well advanced by the end of a few weeks.

Gastric contents

If the last known meal is still present in the stomach of a corpse and the time of that meal is known, then it can give some general indication of the interval between the meal and death. In general if all or almost all of the last meal is present within the stomach then, in the absence of any unusual factors, there is a reasonable medical certainty that death occurred within 3 to 4 hours of eating. Similarly if half of the meal is present then it is reasonably certain that death occurred not less than 1 hour and not more than 10 hours after eating. However, these are broad generalisations and difficulties arise in individual cases because the biology of gastric emptying is complex and influenced by a wide variety of factors including the size and type of meal, drugs, stress and natural disease.

Remarkably liquids, digestible solids and non-digestible solids ingested together in the same meal will leave the stomach at different rates. The emptying of low-calorie liquids is volume- dependant (monexponential) resulting from the motor activity of the proximal stomach. By contrast digestible solids empty more slowly, in an approximately linear pattern after an initial lag period, primarily as a result of the motor activity of the distal stomach. Non-digestible solids which cannot be ground up by the stomach into smaller particles are emptied after the liquid and digestible solids, during the so called inter-digestive period, as a result of a specific wave of motor activity in the stomach. In general meals of a higher osmotic and caloric content are emptied more slowly.

However, there is a substantial variation in gastric emptying rates in normal people. Individuals who suffer severe injuries resulting in coma and survive several days in hospital may still have their last meal within the stomach at autopsy. These are extreme examples of delayed gastric emptying but serve to illustrate the point that the stomach is a poor forensic timekeeper.

There have been several cases of alleged miscarriages of justice in which medical experts have wrongly used the stomach contents at autopsy to provide estimates of time of death to an accuracy of half an hour whereas the degree of accuracy possible is at best within a range of 3 or 4 hours.

Entomology

Insects will colonise a corpse if given the opportunity. The most important flies whose larvae (maggots) feed on corpses belong to the groups Calliphoridae or blow-flies and Sarcophagidae or flesh-flies. The blow-flies are the bright metallic blue and green «bottle flies» commonly found around refuse. Each part of the world has its own indigenous species of these flies and, as a consequence of the movement of human populations, some old world species have been introduced into North America and Australasia. While fly larvae feed on the corpse, beetles feed on the larvae, although some beetle groups will feed directly on the corpse. Beetles appear on the corpse later than flies and are some of the last insects to colonise fragments of soft tissue remaining on skeletonised bodies.

Fly eggs are laid on the moist body parts such as the eyes, nares, mouth, perineum and wounds. Head hair, folds in clothing and the crevice between the body and the ground are sites of oviposition also. Early maggot colonisation of parts of the body not usually colonised suggests that there was a breach in the skin, a wound, at that site to attract oviposition, e. g. on the palms of the hands. After the adult female fly has laid its eggs they hatch within a few hours, depending on species and the ambient temperature, giving rise to the 1st of 3 stages (instars) of larvae. They are very small, usually less than 2 mm in length, and difficult to see. Flesh-flies, unlike blow-flies deposit first instar living larvae rather than eggs on the corpse. First instars moult, shedding their exoskeleton, to produce second instar larvae which grow to a length of up to 4–6 mm. The 2 instars moult to yield 3rd instar larvae, the largest maggot stage, and that most commonly observed; they are voracious feeders on the corpse. When present in large masses they generate considerable heat and a strong odour of ammonia, their main excretory product.

Post feeding larvae, prepupae, migrate from the corpse, wandering off to find a protected place for pupation. The exoskeleton of the 3rd stage larvae hardens and browns forming the puparium. This pupal stage of development is similar to the chrysalis of butterflies where metamorphosis to the adult form of the species occurs. In due course an adult fly will emerge from the pupa.

For each species this life-cycle follows a known temperature-dependant time course. Consequently, maggots of a known stage of development and species found on a corpse give an indication, from the time required for their development, of the minimum period since death. Even in bodies long dead the remnants of insects such as pupa cases and the exoskeletons of beetles may provide useful information.

The pattern of corpse colonisation by successive waves of insects provides a source of further information. Moving a body or burying it some days after death interrupts the normal succession of insects, from which it can be deduced that an event occurred to disturb the normal chain of entomological events. Blow-flies of certain species are found in either an urban or a rural habitat. Finding urban blow-fly larvae on a corpse in a rural setting would suggest death and blow fly oviposition in an urban environment followed by dumping of the body in the rural environment.

The larvae feeding on a corpse may contain any drugs present in the corpse, and are often easier to analyse than body tissue because the corpse contains large numbers of masking chemicals produced by decomposition. Many years after the death, drugs may still be identified in the remnants of pupal cases associated with skeletonised remains.

Botany

Plants and parts of plants may provide evidence of time since death if a plant is in contact with the body or buried with human remains. Ideally a botanist should attend the scene, otherwise colour photographs must be taken and the plant material preserved by drying it between sheets of newspaper. Perennial plants, such as trees, often have seasonal or annual growth rings which can provide a minimum age for human remains where the plant has grown through them or has been damaged by their deposition. Roots can be useful in a similar way.

Annual plants give an indication of time because they complete their life cycles in known time periods in specific seasons, so that disturbances which can be related to a point in the life cycle can be dated. Bodies lying over green plants shade and kill the chlorophyll, and new shoots may develop from damaged stems, changes upon which a time frame can be placed.

SHARP FORCE INJURIES

Sharp force injuries caused by objects which are keen-edged or sharp-pointed are the second major class of injuries. There are two types: incised wounds (cuts) and puncture wounds (stabs). An incised wound is a breach of the skin resulting from contact with a keen edge. Penetration of at least the full thickness of the skin is usual, but shallow, partial-thickness skin cuts can occur, such as the common «paper cut». In clinical practice it sometimes happens that incised wounds and true lacerations are both loosely referred to as lacerations. This may create confusion when clinical medical records are used in legal proceedings because the diagnosis of an incised wound implies contact with a sharp-edged object or weapon, whereas the diagnosis of a laceration implies contact with a blunt object or weapon as the cause of the injury.

A puncture wound or stab wound is a penetrating injury with a depth greater than the wound length on the skin surface. Puncture wounds are caused by objects which are long, thin, and usually sharp-pointed, such as a knife. Both incised wounds and stab wounds typically retain no trace evidence from the weapon, but they are associated with blood staining of the weapon, clothing and scene.

INCISED WOUNDS

Most incised wounds are knife wounds, but this is not a conclusion that can be safely drawn from the examination of the wound alone. Similar wounds are produced by shards of glass, sharp edges of machinery and other sharp-edged objects.

The key features of incised wounds reflect the fact that they are produced by keen-edged objects rather than blunt objects. Incised wounds typically have well defined margins with no associated abrasion or bruising of the adjacent skin and there is always complete severance of all tissues within the depth of the wound, so that there are never any of the tissue bridges which characterise lacerations.

Occasionally incised wounds have irregular or ragged margins because either the causative object had an irregular edge, such as broken glass, or the cut was made across skin folds, such as on the neck or palm, so that the wound assumes a different shape when the folds flatten out. A single incised wound may be discontinuous if it involves surface ridges but misses troughs of soft tissue, for example across the helix of the ear, or across the tip of the nose and adjacent cheek. Similar discontinuous injuries can occur across joints so that the degree of joint flexion at the time of injury can be established by lining up the wounds, as with a cut across the anterior forearm and upper arm sparing the cubital fossa.

Incised wounds inflicted with heavy-bladed instruments, such as an axe, adze, sharpened spade, machete, meat-cleaver, hatchet, sabre, and boat propeller characteristically show abrasion, and sometimes bruising, of the wound margin. This results when the edges of the first-formed incised wound are crushed by the entry of the following thick blade. The prominence of the abrasion is highly variable and usually less prominent if the blade is very sharp and not very thick, such as a sabre. However, the large size of the wounds and the usually associated cuts and fractures to underlying bone are more obvious pointers that the weapon used was heavy-bladed. An associated blow with the blunt poll at the back of an axe blade, or similar weapon may leave an imprint bruise-abrasion.

The shattering of the tempered glass side windows of vehicles during collisions produces a shower of small dice-like fragments which strike the vehicle occupant producing a characteristic pattern of scattered, small, irregular incised wounds associated with small irregular abrasions, the wound shapes reflecting the different angles of impact of the glass fragments. The incised wounds result from the impact of the sharp edges and corners of the dice-like fragments, and the abrasions result from the impact of the flat sides of the fragments. These dicing injuries occur on the side of the face adjacent to the shattered window, so that it is possible to determine whether the individual was seated on the left or right hand side of the vehicle at the time of impact, distinguishing driver from front-seat passenger.

Cuts produced by a serrated knife blade may give an indication of the nature of the blade if the cut is tangential to the skin surface so that the serrations leave shark-fin skin tags at the wound edge, or parallel scratch abrasions on the adjacent skin. Light contact with the skin by serrated blades can produce a line of punctate abrasions or parallel scratches without an associated incised wound. Blade serrations will leave corresponding striations on the cut surfaces of the cartilage of the ribs or larynx.

Accidentally inflicted incised wounds are common. They may result from the use of knifes in the home or workplace, broken window glass in burglaries, broken drinks glasses, and sharp- edged machinery or parts of vehicles.

The ready availability of knifes and the relative painlessness of incised wounds has encouraged their use in suicide since antiquity. The Romans were said to have favoured opening the veins of the arms whilst lying in a warm bath. The Japanese ritual suicide of seppuku involves both self-stabbing and self- incision. Self-inflicted incised wounds are directed towards parts of the body where large blood vessels are close to the skin surface, such as the neck, wrists and, less commonly, the elbows, knees and ankles.

Suicidal incised wounds are commonly, but not invariably, accompanied by parallel, shallow, tentative wounds, reflecting a testing of the weapon as well as the indecision so often present in suicidal acts. These tentative wounds, also referred to as hesitation wounds, show a wide range of appearance between cases. Having produced one deep cut, the person intent on suicide may then produce repeated cuts in the depth of the wound, and these can be seen as multiple trailing cuts arising from the ends and sides of the principal wound.

The stereotypical suicidal cut throat is an incision starting on the side of the neck opposite the dominant cutting hand, and passing slightly obliquely downwards across the midline. Any repeated cuts in the depth of the wound then produce sharply angled tags of skin at the wound ends and edges. The cuts may be so deep that the vertebrae are scored.

Self infliction of shallow cuts as a form of self harm falling short of attempted suicide is seen as parallel, shallowly incised wounds to the fronts of the wrists and forearms which heal leaving multiple, fine, horizontal, linear white scars. Often they are most prominent on the non-dominant arm. When seen in clinical practice, the scars raise the possibility of a personality or psychiatric disorder, with an attendant risk of suicide in custody. If seen at autopsy, they raise the index of suspicion for suicide, alcohol and drug abuse, and high risk behaviour leading to accidental death.

Occasionally individuals self-inflict incised wounds in order to falsely allege that they have been assaulted. These factitious injuries have a pattern which reflects handedness, easy accessibility of the injured parts, infliction by pressing the sharp edge against the skin and then running it across the skin surface rather than slashing movements, and the avoidance of sensitive or critical areas such as nipples, lips and eyes. Although usually superficial, these factitious injuries can be extensive and mutilating. Some individuals have self-inflicted such injuries in the pattern of the stigmata of Christ, claiming that they appeared spontaneously.

Individuals defending themselves against a knife attack commonly sustain defensive injuries to the hands and forearms, which range from subtle to devastating. Defensive wounds to the legs may be present in a victim who was attacked when on the ground. Cuts to the palms and fingers result from attempts to grab or deflect the weapon, while slash and stab wounds to the backs of the hands and the forearms result from shielding movements. By contrast with the pattern seen in defensive wounds, if the knife slips within the hand of the assailant then the blade typically cuts either the base of the little finger or the web of the thumb of the assailant.

Defensive injuries to the hands and forearms although typical of knife attacks may be absent in sudden, overwhelming attacks (so called «blitz» attacks), or if the victim is unable to offer a defence as a result of the effects of alcohol and drugs, unconsciousness, bindings, or other physical and emotional circumstances. Since defensive wounds reflect anticipation of injury and an attempt to ward off the harm, they may be seen in accidents as well as assaults. For example, multiple cuts to the palms of the hands can be produced in a fall onto a glass-strewn floor, as well as during an attack with a broken bottle.

Assaultive incised wounds to the face may be inflicted with the intention to disfigure. In a prison setting the classical improvised weapon for this purpose is a plastic tooth-brush handle into which is embedded two closely placed parallel razor blades. The Stanley knife, a work knife which has a very sharp retractable blade, is another weapon commonly used.

Cleanly severed blood vessels within an incised wound bleed profusely, and if large vessels are cut then the haemorrhage may be sufficient to kill. Such extensive haemorrhage commonly results in blood soaking of clothing together with blood staining and spattering of surroundings, opening up the possibility of reconstruction of the events through blood spatter analysis. When large veins are severed, particularly in the neck, air may be drawn into the circulation, obstructing the flow of blood through the heart and killing by air embolism

STAB WOUNDS

Stab wounds, like incised wounds, typically yield no trace evidence, are associated with blood and tissue staining of the weapon, as well as blood staining and damage to clothing, and offer the opportunity for reconstruction of the events from interpretation of blood spatter at the scene.

Stab wounds are penetrating injuries produced by a long, thin object, which is typically pointed. Most commonly the instrument is flat with a sharp point, such as a knife, a shard of glass or a length of metal or wood. Other weapons may be long and thin with a sharp point, such as a skewer, hypodermic needle, ice pick, or old-fashioned hatpin. With sufficient force even long, rigid objects which are blunt-ended will produce puncture wounds, e. g. a screwdriver, wooden stake, or protruding parts of machinery or motor vehicles. The appearance and dimensions of the resulting wound often provides useful information about the object which produced it.

The skin surface appearance of a stab wound is influenced both by the nature of the weapon and characteristics of the skin. The skin contains a large amount of elastic tissue and will both stretch and recoil. This elastic tissue is not randomly distributed but is aligned so as to produce natural lines of tension (Langer's lines) which can be mapped out on the skin surface. In survivors of knife assaults, the extent of wound scarring will be influenced by the alignment of the wounds relative to Langer's lines.

Wounds which have a long axis parallel with Langer's lines gape only slightly, a fact made use of by surgeons who align their incisions in this way to promote healing and reduce scarring. Wounds aligned at right angles to Langer's lines tend to gape widely, and scar prominently, because the natural lines of tension of the skin pull the wound open. Wound gaping is also influenced by the extent of damage to the underlying supporting fascia and muscles.

When a stab wound in a corpse is gaping then the wound edges must be re-approximated to reconstruct the original shape of the wound, something that is easily achieved with transparent tape. It is the dimensions of the reconstructed wound rather than the original gaping wound that are of value in predicting the dimensions of the blade. One purpose in examining stab wounds is to establish whether potential weapons could have or could not have produced the wounds.

The dimensions of a stab wound give some indication of the dimensions of the blade of the weapon. If a stabbing with a knife is straight in and out then the length of the stab wound on the skin surface will reflect the width of the knife blade. However, there are important qualifications which apply. The skin wound length may be marginally (a few millimetres) shorter than the blade width as a result of the elastic recoil of the skin. If the knife blade has a marked taper and the entire length of the blade did not enter the body then the skin wound length may not represent the maximum width of the blade. If the blade did not pass straight in and out but instead there was some rocking of the blade, or if it was withdrawn at a different angle from the original thrust, then the skin wound will be longer than the inserted blade width.

Consequently, the most reliable assessment of blade width is made from the deepest wound with the shortest skin surface length, because this wound represents the greatest blade penetration with least lateral movement. A single weapon can produce a series of wounds encompassing a wide range of skin-surface lengths and wound depths. This is often seen in a multiple stabbing fatality and is consistent with the use of only one weapon. However, it is rarely possible to exclude any speculative suggestion of more than one weapon, and, by inference, more than one assailant.

The depth of the wound gives an indication of the length of the weapon. Clearly the wound track length may be less than the blade length if the entire blade did not enter the body. Less obvious is the fact that the wound track length may be greater than the blade length. This may occur if the knife thrust is forceful and the tissues are compressed, so that when the weapon is withdrawn the track length in the now decompressed tissues is greater than the blade length. This tissue compression effect is most marked in wounds to the anterior chest and abdomen, since the entire chest or abdominal wall can be driven backwards by the blow.

A small pocket-knife, with a blade of about 2 inches or 5 cm, can cause, in a slim person, a fatal stab wound to the heart or one which penetrates the abdomen to transfix the aorta. An added difficulty in measuring the wound track length at autopsy is that the corpse is supine with the viscera in a slightly different relative position to a living person standing or sitting, and with the tissues dissected and displaced. For all of these reasons the wound track depth should be used with caution in predicting the blade length of the weapon. If by chance some fixed bone, such as a vertebra, is damaged at the end of the wound track, then the assessment of depth of penetration is easier, but still subject to inaccuracy.

As well as providing an indication of blade width and blade length, a stab wound may provide other useful information about the weapon. Wound breadth on the skin surface is a reflection of blade thickness and a typical small kitchen knife with a blade thickness between 1 and 2 mm produces a very narrow wound. The use of a thicker-bladed weapon may be readily apparent from the measured wound breadth on the skin surface.

Most knives have a single-edged blade, with one keen edge and one blunt edge to the blade. The resultant wound reflects the cross-sectional shape of the blade and, with the wound gaping, often appears boat-like with a pointed prow and a blunted stern. Skin elasticity may distort the blunted stern shape into a double-pronged fishtail. The thicker the blade of the weapon, the more obvious is the blunting of one end of the wound when contrasted with the other pointed end. Knives with double-edged blades (daggers) are specifically designed for use as weapons and produce a wound that is pointed at both ends, but an identical double-pointed wound can be produced by a thin, single-edged blade.

If the blade penetrates to its full length then the wound may show features from the hilt and adjacent part of the blade. A knife with a folding blade has a notch at the hilt end of the sharp edge of the blade. This notch, or «kick of the tang» (the tang being the rectangular portion of metal from which the blade was forged), leaves a rectangular or square abrasion at the end of the wound if the blade penetrates to its full length. A hunting-type knife, having a rectangular metal plate between the blade and the handle, leaves rectangular abrasions, which may be subtle, at both ends of the wound. Any imprint from the hilt of a weapon is usually only partial because the knife entry is most often at an angle. Dove-tail abrasions originating from either side of the stab wound are typical of a wooden-handled blade.

A semi-lunar abrasion at one side of a stab wound is typical of a pen-knife handle. Other protrusions from the base of the blade or hilt may leave abrasions characteristic of the weapon. Serrated blades or blades with a saw-type component, as is seen in some hunting knives, can leave parallel linear abrasions or incised wounds on the skin in association with the stab wounds. Serrated blades leave corresponding striations on cartilage and occasionally the dermis. Mottled bruising associated with the wound may be produced by the knuckles of the assailant's hand, and an imprint bruise-abrasion from the clothing reflects a similar mechanism of forceful impact from the hand.

The cross-sectional shape of the weapon, particularly the back of the blade, may be accurately reproduced if it passes through bone, e. g. skull, pelvis, sternum or ribs. At the same time, passage through bone may cause trace material from the blade, e. g. paint, to be scraped off and deposited in the wound. Stab wounds in solid organs such as the liver retain the profile of the weapon, and this can be visualised by filling the wound track with a radio-opaque contrast material and taking an x-ray, or by making a cast of the wound using modern dental casting material.

Weapons other than knives may produce characteristic stab wounds. Bayonets, which have a ridge along the back of the blade and a groove along either side, to lighten the blade and facilitate withdrawal of the weapon, produce wounds like an elongated letter «T». A pointed metal bar which is square in cross-section typically produces a cruciate wound, whereas one which is circular in cross-section, e. g. a pitchfork, produces an elliptical wound. Similarly a triangular file will produce a 3-cornered wound. If the cross-sectional shape of the weapon varies along its length, e. g. a screwdriver, then the depth of penetration will affect the appearance of the wound. A forked instrument produces paired stab holes at different distances depending upon the angle of penetration. A scissors leaves paired wounds, off-set and pointed towards each other. A screwdriver or arrow wound may be indistinguishable from a gunshot wound because all leave a central defect with a marginal abrasion produced by the same mechanism of an inward pushing with grazing of the skin margin. Post-mortem drying turns the marginal abrasion dark brown or black.

Relatively blunt instruments such as pokers, closed scissors and files, tend to bruise and abrade the wound margins, a feature not otherwise seen in stab wounds. The blunter the object and the thicker its shaft then the more likely will the skin surface wound become a ragged, often cruciate, split. In cases where the wound appearance is unusual, it is helpful to conduct experiments with a duplicate suspect weapon and pig skin, in order to see whether the wound appearances can be reproduced.

Stab wounds inflicted during a struggle, with knife thrusts at awkward angles and with movement of both victim and assailant during the attack, may show characteristics reflecting this. Even so, it is rarely if ever possible to reconstruct the positions of victim and assailant from the location and direction of the wounds. A notch on the otherwise cleanly cut edge of the wound is a result of withdrawal of the blade at a different angle from the entry thrust. Exaggeration of this effect leads to a V-shaped or even cruciate wound when there is marked twisting of the blade or twisting of the body of the victim. By contrast, clothing tends to twist with the blade so that a linear stab hole in clothing may overlie a V-shaped stab in skin. A linear abrasion (scratch) extending from one end of the wound results from the withdrawn blade tip running across the skin. A single stab hole on the skin surface may be associated with more than one wound track through the tissues, reflecting a knife thrust followed by continuing struggle or repeated thrusts of the weapon without complete withdrawal.

A commonly asked question in the courts is the amount of force required to produce a specific stab wound. This is usually a difficult if not impossible question to answer. The sharpness of the point of the weapon is the most important factor in determining the degree of force required to produce a stab wound. In general, relatively little force is required to produce a deeply penetrating stab wound using a sharply pointed weapon, and the amount of force is easily over¬estimated. The greatest resistance to penetration is provided by the skin and once this resistance is overcome then the blade enters the tissues with greater ease. In this respect an analogy can be made with the stabbing of a ripe melon.

The important implication is that the depth of the wound is not a measure of the degree of force applied. However, penetration of any bone implies a significant degree of force; all the more so if the tip of the blade has broken off and remains embedded in the bone, something which is best identified by X-ray. Similarly a significant degree of force may be inferred from the presence of the hilt mark of the weapon on the skin surface, an uncommon finding, or a wound track significantly longer than blade length, suggesting forceful tissue compression during the stabbing. Even so, the stabbing force may have been a combination of both the thrust of the weapon and also any forward movement of the victim, such as in a fall. This latter proposition is commonly raised by the defence, and is rarely possible to discount, in deaths from single stabs.

BLUNT FORCE INJURIES

A wound is a disruption of the continuity of tissues produced by external mechanical force. The term injury is used synonymously with wound, but can have a wider meaning and encompass not only damage produced by physical force, but also damage produced by heat, cold, chemicals, electricity and radiation. The English word «injury» derives from the Latin word «injuria» which literally means not lawful. The term lesion originally meant an injury but has now come to be more widely applied to include any area of injury, disease, or local degeneration in a tissue causing a change in its function or structure. Therefore use of the terms injury or wound imply damage from unnatural causes, while the use of the term lesion is non-committal on whether the cause of the damage was natural disease or not.

Wounds, or injuries, are generally classified according to their cause as blunt force injuries, sharp force injuries, gunshot injuries, and a miscellany of others, including for example burns. This is the classification followed in this text, and each group of injuries is discussed in successive chapters.

A blunt force injury is any bodily damage resulting from forceful contact between the body and a blunt object. The forceful contact most commonly involves movement and impact, with the resulting transfer of kinetic energy. Either the moving object strikes the body as in a blow, or the moving body strikes an unyielding object as in a fall. Less often the physical force is applied more slowly by the pressure of crushing, squeezing, or pinching. Crushing or scraping of the skin produces abrasions (grazes). Bruises occur when the elastic limit, the tolerance, of subcutaneous blood vessels is exceeded so that they tear and bleed. Greater forces are needed to tear the skin to produce lacerations. These three blunt force injuries — abrasions, bruises, and lacerations — may occur singly or together in any combination.

ABRASIONS

An abrasion is an area of crushing or loss of skin, or mucous membrane, resulting from contact with a blunt object. Abrasions are typically superficial, trivial injuries which may be overlooked easily but provide useful forensic information. They bleed only slightly, heal quickly and leave no scar. Tangential impact between an object and the skin causes a typical graze in which the superficial skin layers are scraped off and piled up as skin tags at the far end of the injury. The location of the skin tags indicates the relative direction of movement between the object and the body.

Broad patches of abrasion are sometimes described as brush abrasions, the frictional element of which gave rise to the term «brush burns», as in for example «carpet burns». They are seen in their most florid form as «road rash», following a motor vehicle collision in which the victim, typically a motorcyclist, slides along the road surface.

Trace evidence from the impacting surface may be present within a graze and can be recovered for identification and matching with the impacting object, for example paint from a vehicle in a hit and run death of a pedestrian. Similarly, fragments of skin transferred to the impacting object may be matched to the victim using serology or DNA techniques. Examination of a graze for the presence of skin tags and trace evidence is easier using a hand-held magnifying glass.

A scratch is a linear abrasion produced by drawing a sharp point over the surface of the skin or mucous membrane. Plant thorns and barbed wire are common causes of scratches. The presence of skin tags at one end indicates the direction of the scratch. Human fingernails and animal claws are other important causes of scratch abrasions. The spacing and convergence of several scratch marks may allow a distinction between human and animal and between adult and child. Small quantities of skin can be recovered from beneath fingernails which have been used for scratching, and linked to the person who was scratched by serology or DNA techniques. Tissue from beneath the fingernails is recovered by clipping the nails with clean scissors, scraping beneath the nails using wooden «orange sticks», or cleaning beneath the nails using pointed cotton swabs.

Whereas tangential impact between an object and the skin will produce a graze or scratch, direct impact at approximately right angles (90°) to the skin surface will produce a crush abrasion, which imprints the pattern of the causative object upon the skin surface. These imprint or patterned abrasions may result from forceful impact, as in a pedestrian struck by a vehicle, or alternatively may result from a more slowly applied crushing force, such as in strangulation by ligature. The importance of patterned abrasions lies in the fact that their shape can be matched to the causative object, and for this reason they should be documented by photography with a scale, or by an accurate tracing.

Common examples of imprint abrasions are the sole patterns of footwear in kicking assaults, and motor vehicle tyre tread marks in run-over pedestrian deaths. In deaths from hanging the ligature typically leaves a distinct imprint abrasion which is accentuated by the continued suspension of the body after death. Fingernails, as well as producing scratch abrasions, can produce semi-lunar and linear imprint abrasions. The hilt of a knife may leave an imprint abrasion adjacent to a stab wound, suggesting the forceful penetration of the entire length of the blade. A diagonal abrasion across the chest of a motor vehicle collision victim suggests that a three-point seatbelt was being worn at the time of impact.

In the most superficial abrasions there is crushing and loss of only the epidermis so that there is no bleeding and often no inflammatory reaction. Slightly deeper abrasions involve the superficial dermis, appear red-based, may bleed slightly, and produce an exudate which dries within a day to form a yellow- brown scab. Abrasions produced after death, or immediately prior to death, have a dried red, orange or yellow base and are described as parchmented abrasions because they resemble parchment. As the post-mortem interval increases these parchmented abrasions dry out and harden further, becoming a dark brown or black colour. Superficial abrasions on a wet or damp corpse may escape notice until they dry out and darken.

Overall, abrasions provide useful forensic information because their location reflects precisely the point of impact with the causative object, grazes indicate direction, imprint abrasions reflect the shape of the causative object, trace evidence from the object may be found in the graze, and skin fragments from the victim may be recovered from the causative object.

Bruises

A bruise is a haemorrhage into tissues. Synonyms are contusion and ecchymosis (plural: ecchymoses). Very small bruises, ranging in size from a pinpoint to a pinhead, may be described as petechiae, or petechial haemorrhages, or alternatively as punctate haemorrhages. Haemorrhage or bleeding is the process which produces a bruise in tissues, but the term haemorrhage also encompasses bleeding which may not be associated with bruising, e. g. a bleeding nose, or a bleeding stomach ulcer.

Any tissue may bruise, but bruises confined to deeper tissues, such as skeletal muscle, are not visible on the skin surface. Bruises of the deep tissues, even when fatal, may not be evidenced by any injury to the skin surface. For example, a fatal head injury, such as a sub-dural haemotoma, may be encountered without recognizable superficial bruising; fatal strangulation with extensive bruising of the muscles of the neck may be accomplished without obvious bruising of the skin; and blows to the abdomen, although producing bruising and ruptures of internal organs, may not produce any abdominal wall bruising.

Only bruising of the skin and subcutaneous tissues, and the mucosa of the mouth, vagina and anus, is visible at a clinical examination. A simple bruise is a discolouration resulting from haemorrhage beneath the skin or mucosa, without any associated breach in the surface. The blood vessels ruptured are typically the capillaries and small veins rather than arteries. Skin and mucosal bruises may be accompanied by abrasions and lacerations, but they are not usually associated with cuts and stabs, where there is a free flow of blood from the cut vessels rather than infiltration of blood into the tissues. Bruises are accompanied by swelling from the haemorrhage itself and the resulting inflammation.

If the extravasated blood collects as a discreet tumour-like pool, the lesion is referred to as a haematoma. Bruises are usually painful and tender to palpation as a result of damage to local nerve endings and the inflammation. Focal necrosis of subcutaneous fat may occur at the site of a bruise, and a secondary aseptic inflammation in response to the irritant effect of fat liberated from the ruptured cells produces a hard chronic lesion. This is more of clinical than forensic importance since a common site is the breast where it may be mistaken for a carcinoma. Bruises of the skin if numerous and large enough can be life-threatening as a result of blood loss but this is an uncommon occurrence.

Bruises are produced typically by a blunt force impact, such as a blow or a fall, but can result also from crushing, squeezing or pinching. Bruising, with or without abrasion, to the bony prominences of the back (i. e. the shoulder blades, sacrum and pelvis) may be caused by force applied to the front of a supine body with resultant counter-pressure between an underlying firm surface and the back, as in forceful restraint on the ground during a rape, or a stamping assault.

Similar «counter- pressure bruising» may be seen on the bony prominences of the front of the pelvis (the anterior superior iliac crests) in attempted anal rape. So called «love-bites» («hickeys» in American slang) are superficial bruises produced by the negative pressure of mouth suction. They are commonly found on the side of the neck and occasionally the breast and inner thigh.

They are sometimes self-inflicted on accessible parts of the arms to simulate evidence of an assault. Natural diseases in which there is an abnormality of the clotting mechanism of the blood, such as leukaemia, scurvy (vitamin C deficiency), and liver disease, cause so-called 'spontaneous bruising' which is thought to result from unrecalled trivial trauma. Florid spontaneous bruising (purpura) may be seen in children with fulminating meningococcal infection.

The size of a bruise is an unreliable indicator of the degree of force which caused it, because several other factors, such as anatomical site, gender, age and the presence of natural disease, all influence bruise size. Bruising occurs more readily where there is a lot of subcutaneous fat, such as on the buttocks and thighs, and therefore more readily in the obese, in women and in infants. Bruising occurs readily in loose tissues, such as around the eyes and genitals, and less readily where the skin is strongly supported by fibrous tissue, such as the palms, soles and scalp. The elderly bruise more easily because of degenerative changes in the blood vessels and the supportive tissues of the skin and subcutaneous fat.

Senile purpura, characterised by sharply-defined geographic areas of «spontaneous» bruising to the backs of the hands and forearms is found in the very old or frail. Natural diseases which affect blood clotting, degenerative diseases of small blood vessels, and high blood pressure make some individuals more susceptible to bruising. Skin colour does not change the extent of bruising but does significantly influence its appearance because the bruise is viewed through the semi-translucent skin. Bruising is most easily seen in very pale-skinned individuals, particularly blondes and redheads, and can be completely masked by the natural skin colour of blacks. Examination under ultraviolet light helps reveal bruises which are otherwise difficult to see. Black and white photography, which is more sensitive to ultraviolet light than colour photography, may allow the bruise to stand out more clearly in a photograph.

Bruises tend not to reflect accurately the shape of the object which produced them, and they change shape with time. Exceptionally the surface detail of the striking object may be imprinted as a patterned bruise on the skin, often associated with a patterned abrasion. Such patterned or imprint bruises typically occur following a heavy impact, such as from a shod foot or motor vehicle, or from the muzzle or foresight of a gun in a contact gunshot wound, with death occurring soon after injury, and so limiting the diffusion of blood and the obscuring of the imprinted pattern. Patterned bruises commonly have a bright red intra-dermal component whose diffusion is limited also by the dense collagenous dermal tissue. Occasionally clothing or jewellery may leave a patterned bruise on a body when it is crushed into the skin surface by an impacting object, e. g. a motor vehicle striking a pedestrian, or a kick through clothing.

Sometimes bruises give a more general impression of the causative object, for example a doughnut-shaped bruise is produced by a hard object with a rounded contour, such as a cricket ball or a baseball. A similar mechanism produces two parallel linear bruises, so-called «tram-line» or «rail-track» bruises, as the result of a blow from a long object which has a circular cross-sectional shape, such as a police baton, or an electric flex. If the flex has been looped then this may be apparent as curved rail-track bruising.

When a blow with a rod is struck against the buttocks, — a particularly pliable, curved, soft surface — the tissues are compressed and flattened under the impact and the resulting rail-track bruise will follow the curved contour of the buttocks. A pliable weapon such as a strap or electric flex may produce a similar appearance because it can wrap around the body on impact. Bruises produced by finger-pads as a result of gripping are usually larger than the finger-pads themselves, but their number, pattern and location on the victim suggests the mechanism of causation. Finger-pad bruising is seen on the neck in throttling, on the upper arms in restraint, on the thighs in rape, and on the chest and face in child abuse.

Bruises change colour as they age before finally fading away. A fresh bruise is dark red, the colour of venous blood, turning soon to a dusky purple. Thereafter the colour changes progressively from the periphery of the bruise towards the centre through brown, green, yellow and a pale straw colour before disappearing. These colour changes reflect the breakdown of haemoglobin into coloured products as part of the inflammatory process. The time frame of the colour change is extremely variable depending upon bruise size, depth, location and the general health of the individual, but most bruises disappear within one to four weeks. «Love bites», which are small and superficial, typically complete this sequence in 7 days.

In general, bruises which have a green or yellow margin are 3 or more days old and those which appear entirely dark red or dusky purple are fresh having occurred within a day or so. The accurate estimation of the age of a single bruise is not possible but fresh bruises are easily distinguished from bruises several days old. Establishing that bruises are of different ages, and therefore inflicted at different times, is important in the assessment of allegations of repeated assaults, such as in child abuse and spousal abuse. Chronic alcoholics commonly have multiple bruises of different ages over their legs and arms as a result of repeated falls when drunk, often with the extent of bruising made worse by disturbances of blood clotting secondary to alcoholic liver disease.

The colour of bruises does not change after death, but they may become more evident against the now pale skin, or alternatively be obscured by the post-mortem skin colour changes of lividity and decomposition. It is not possible to distinguish a bruise sustained at the time of death from one which occurred up to a few hours earlier, and such bruises are best described as having occurred «at or about the time of death». If the microscopic examination of a bruise shows an inflammatory reaction then it was likely inflicted more than a few hours before death, and almost certainly more than half an hour before.

Haemoglobin and bilins

Haemoglobin, the oxygen-carrier of red blood cells, is a conjugated protein in which globin, the protein component, is combined with haemin, a ferric porphyrin component. The intense colour of haemoglobin is not caused by iron but by porphyrin, a complex cyclic structure of four pyrrole rings (each pyrrole ring is a 5-membered closed structure containing 1 nitrogen and 4 carbon atoms) linked to each other by methine groups (-CH=). The iron atom is kept in the centre of the porphyrin ring by interaction with the four nitrogen atoms.

In oxyhaemoglobin, which is bright red, the ferrous ion is bound to oxygen. Deoxyhaemo-globin, or reduced haemoglobin, is oxyhaemo-globin minus oxygen and is purple in colour. The metabolic products of the haeme portion of haemoglobin include a series of yellow, green, red, and brown non-metallic compounds arranged as linear chain structures of pyrrole rings rather than in the cyclic configuration of porphyrins. These are the bilins, or bilichromes. One of these bilins is the blue-green compound, biliverdin (C_{33} H_{34} O_6 N_4), and another, bilirubin (C_{33} H_{36} O_6 N_4), is a yellow-red product formed from biliverdin by the addition of 2 hydrogen atoms. It is these and other biological pigments, or biochromes, formed as catabolic products of the porphyrin of haeme which are responsible for the colour changes in bruises.

Bruises to the deep tissues can be present without any evident skin surface injury, particularly if the force applied is by a smooth object over a wide area. Such deep bruises may spread under the influence of gravity and body movement, following the path of least resistance along natural or traumatic planes of cleavage of the tissues. This shifting of deep bruises explains their delayed appearance at the skin surface some days after infliction, often at sites distant from the points of impact. The delayed appearance of bruising around the eyes follows a blow to the forehead, bruising behind the knee follows a blow to the lateral thigh or a fractured neck of femur, and bruising to the neck follows a fractured jaw. A 2nd examination of a victim of an assault after an interval of a few days may show visible bruising where previously there had been only the swelling or tenderness of deep bruising. Such 2nd examinations are recommended as best practice.

Bruising is essentially a vital phenomenon in which the infiltration of blood into the tissues occurs under the pressure of the circulating blood. After death, the lack of blood pressure means that it requires considerable force to produce a bruise in a corpse. Such post mortem bruises are disproportionately small relative to the force applied, which may be evident from associated fractures, and the resultant bruises are usually only a few centimetres in diameter. In assessing the possibility that bruising may be post mortem, the findings and circumstances as a whole should be considered, and against this background quantitative differences between ante mortem and post mortem bruises are usually so great that confusion is unlikely. It is seldom difficult to distinguish between injuries with vital bruising resulting from a vehicle running over a live body, and the tearing and crushing of dead tissues. However, finger-pad bruises to the insides of the upper arms may be produced by simply lifting a corpse, particularly in those elderly women who have abundant loose upper-arm fat, which is often congested due to post mortem lividity. Any livid dependent areas of a corpse bruise more readily post-mortem as a result of the vascular congestion.

Post-mortem lividity, which is gravitational pooling of blood within the blood vessels after death, may be confused with bruising. However, the pattern and distribution of lividity usually makes the distinction straightforward. In doubtful cases, incision of the skin discloses blood oozing from the cut ends of vessels and washing the cut surface removes the blood, whereas the blood infiltrating the tissues in bruises cannot be washed away. If needed the distinction may be confirmed by microscopic examination.

In a fair-skinned corpse the congested muscles of the base of the thumb (thenar eminence) and dorsum of the foot may be visible through the thin overlying skin and superficially resemble bruising. Post mortem decomposition with its initial green discolouration of the anterior abdominal wall is readily distinguished from bruising by its colour and location. Putrefactive haemolysis of blood within blood vessels and decompositional breakdown of vessel walls results in extravasation and diffusion of haemolysed blood into adjacent tissues, and this haemolytic staining of tissue may entirely mask small ante mortem bruises, e. g. in the neck muscles in cases of throttling.

Lacerations

A laceration is a tear in tissue. Any tissue may be lacerated, and for example a fracture of bone or a rupture of an Achilles tendon is a laceration. Only lacerations of the skin are discussed here, and lacerations of deeper tissues are discussed in later chapters. Split-like lacerations of the skin are the result of direct crushing, typically between an unyielding object on the one side and underlying bone on the other. Skin splits occur most commonly where the skin lies close to the underlying bones, such as the scalp, eyebrows, cheeks, elbows, knuckles, knees, and shins.

The splitting or tearing of the skin which produces a laceration characteristically leaves intact some strands of stronger subcutaneous tissue, such as small arteries, nerves and fibrous bands, which straddle the depths of the wound. These tissue bridges are of diagnostic importance in identifying the injury as a laceration and therefore as a result of blunt force, and not a cut wound resulting from a sharp-edged object. Lacerations are ragged-edged and commonly associated with bruising and abrasion, the latter indicating the point of impact. However, some lacerations produced by impact against a thin object, or edge of an object, can have sharply defined margins with no apparent bruising or abrasion, so that they may be confused with cut wounds. Inspection of the wound margins using a magnifying glass often allows the distinction to be made between a laceration and a cut, and recognizing tissue bridges within the depth of the wound establishes conclusively that the wound is a laceration.

Lacerations do not generally reflect the shape or size of the object which produced them; they may be smaller than the impact area or may extend beyond the impact area. A stellate laceration radiating from a contact area follows a blow from an object such as a poker end. A linear weapon, such as a stick, commonly causes a linear laceration which may bifurcate at one end giving it a Y-shape, or bifurcate at both ends. If the impact is tangential and raises a skin flap then the direction of the flap is the direction in which the force was applied. A hammer, or weapon with a similar circular striking face, produces circular or crescentic lacerations. Lacerations to the insides of the lips result from crushing of the lips against the teeth following a blow to the mouth. Sometimes these mucosal lacerations can be matched to the edges of the teeth, particularly teeth which are mal-aligned and protruding or sharp-edged from caries.

Lacerations are produced also when the skin is stretched and the forces exceed the elastic limit, or tolerance, of the skin. These tears or stretch lacerations are seen in the groins of pedestrians struck from behind by a vehicle, or of persons who have fallen from a great height and landed with the legs bent back on themselves. Over-stretching injuries also occur in falls from aeroplanes or sky-scrapers as a result of flailing of the limbs, a phenomenon known as «wind-milling». Extreme tangential shearing forces, such as those produced by motor vehicle tyres when a pedestrian is over-run, produce flaying (or «degloving») injuries with the raising of very large areas of skin from the underlying soft tissues.

Lacerations penetrate the full thickness of the skin to involve the underlying tissues and will bleed, but less so than a comparable cut wound. The disrupted blood vessels and tissue in a laceration are partly crushed and initiate thrombosis by the extrinsic pathway so reducing blood loss. The depths of a laceration often contain foreign material, a source of wound contamination to be cleaned away in clinical practice, but a source of trace evidence to be collected in forensic practice. Trace evidence recovered from a laceration may link it to the impacting object, while blood and tissue recovered from the suspect object may be linked to the injured victim, through DNA techniques or less sophisticated blood and tissue typing.

GUNSHOT INJURIES

An understanding of gunshot wounds requires some knowledge of the construction and mechanism of guns and the ammunition which they fire. Rifles and handguns, but not shotguns, have rifled barrels. The rifling consists of spiral grooves cut the length of the interior or bore of the barrel. When a rifled weapon is discharged, the rifling causes the bullet to spin around its long axis, like a spinning-top, and this lends gyroscopic stability to the bullet in flight, increasing the accuracy of the weapon. Shotguns by contrast have a smooth bore with no rifling and are designed to fire multiple pellets, although they can fire a single slug. Rifled weapons, their ammunition and the injuries they produce are discussed first. Smooth bore weapons, or shotguns, their ammunition and the injuries which they produce are discussed later.

RIFLED WEAPONS

Handguns and rifles are the 2 most frequently encountered rifled weapons. Handguns, as the name implies, are designed to be fired from the hand and may be revolvers or auto-loading pistols, so called «automatics». Revolvers have a revolving cylinder containing several chambers each of which holds one cartridge, whereas auto-loading pistols have a removable magazine storing cartridges with a mechanism for autoloading after each firing. Revolvers do not have manual safety catches. Rifles are firearms with a rifled barrel designed to be fired from the shoulder, and there are many designs around this theme. Other rifled weapons used by the military include submachine guns, which are designed to be fired from the shoulder or the hip and are capable of fully automatic fire, and machine guns which are heavier weapons also capable of fully automatic fire.

The ammunition or cartridges, for these rifled weapons consists of a cartridge case within the base of which is a chemical primer, with a chemical propellant above and the bullet clamped on top. The word bullet comes from the French word boulette which means a small ball, reflecting the fact that original projectiles were small lead spheres, the musket ball. Rifles are either rim-fire, having the primer located in the rim of the cartridge case, or are centre-fire, having the primer located at the centre of the cartridge case. Rim-fire rifles produce wounds similar to those of low velocity handgun wounds. Rim-fire cartridges are used mostly in small-bore, low velocity weapons, both handguns and rifles, of 22 calibre.

When the trigger of the gun is pulled, the firing pin strikes the base of the cartridge case where the primer is located, causing the primer to explode and to ignite the propellant. The propellant burns to produce large volumes of gasses under pressure. The original propellant, or gunpowder, used in firearms was black powder but this is now obsolete and smokeless powder is used instead. The heat generated by the burning propellant causes the cartridge case, which is typically made of brass, to expand and seal the chamber against any rearward escape of gasses. The large volume of hot gasses forces the bullet out of the cartridge case and down the rifled barrel to exit the muzzle of the weapon. Bullets come in a variety of shapes. The simplest are made from lead alloys and are traditionally fired from revolvers and 22 inch calibre rifles. Other bullets have a lead core encased by a copper alloy metal jacket. In military ammunition this metal jacket completely encases the lead core — full metal jacketing. In hunting rifle and semi-automatic pistol ammunition the lead core is partly exposed at the tip of the bullet — partial metal jacketing.

Wound ballistics

Ballistics is the science of the motion of projectiles and wound ballistics is the study of the projectile penetration of tissues. A moving projectile, or bullet, has kinetic energy proportional to its weight and its velocity squared (KE a WV^2). The wounding potential of a bullet is directly related to its kinetic energy and since kinetic energy increases in proportion to increases in the velocity squared, it is high velocity projectiles which have the greatest wounding potential. The transfer of kinetic energy from the bullet to the tissues produces the wounding effect. If the bullet does not exit the body then all its kinetic energy will be transferred to the tissues, but if it exits the body then only some of its kinetic energy will have been transferred. Wobbling and tumbling of the bullet as it passes through the body, bullet deformation and break-up, long wound tracks and passage through denser tissues are all factors which influence the loss of kinetic energy and increase the wounding effect.

Bullets disrupt tissues by 2 principal mechanisms. Direct laceration of the tissues occurs as the bullet penetrates just as with any penetrating object. This is the main mechanism of tissue damage in low velocity gunshot wounds from such weapons as pistols, and the permanent cavity visible in the body accurately reflects the tissue damage produced by the bullet. The second mechanism of tissue damage is temporary cavity formation and occurs only in high velocity rifle wounds. The transfer of kinetic energy from these bullets is go great that they create a temporary cavity up to 30 times the diameter of the bullet at a pressure of 100 to 200 atmospheres over a time span of 5–10 milliseconds.

The result is a small permanent cavity surrounded by a wide zone of haemorrhage and an even wider zone of tissue damage which may not be obvious on examination. Other mechanisms of tissue damage include shock-waves, travelling at the speed of sound, which may cause the rupture of gas-filled organs such as the bowel; the creation of secondary projectiles such as shattered bone fragments in gunshot wounds to the head; and the effects of discharge gases which enter the wound track and produce tissue disruption if there is hard contact between the muzzle of the weapon and the body at the time of firing.

Rifled gunshot wounds

A gunshot wound is a penetrating wound which leaves a skin defect where the projectile passes through the skin. As a general rule a gunshot exit wound is larger and more irregular than a gunshot entry wound due to the effect of bullet tumbling and bullet deformation. Characteristic of entry wounds, particularly from handguns, is an abraded margin which surrounds the skin defect and is a result of the forward motion of the bullet indenting the skin and grazing it at the time of entry.

The abrasion ring appears reddish-brown but can darken to almost black with post mortem drying. An abraded margin is typically absent from exit wounds, aiding their distinction from entry wounds. However, an abraded margin may be seen occasionally around an exit wound when the skin has been «shored up» by a firm surface, such as a wall or a belt, so that the everted skin margins of the exit wound are impacted against the supporting surface. High velocity rifle bullets tend not to produce an abraded margin but rather micro-tears, which are fine radiating lacerations around the edge of the skin defect, best seen with a magnifying glass.

Lubricant and debris on the bullet surface is wiped off onto the wound edge and appears as a grey ring of bullet wipe, but this is often obscured by blood and is more frequently seen on overlying clothing. The presence of bullet wipe is an indication that the wound is an entry and not an exit wound. As well as the bullet, other products of discharge of the weapon exit the muzzle and may deposit on the clothing, the skin or enter the wound track. These products of discharge include soot, unburned, partially burnt and burning grains of propellant, and hot gasses including flame, which is incandescent gas. Together the effects of these are sometimes termed «powder burns», but the more precise terms which refer to the effects of each element are preferable. Deposition of soot is described as smudging; propellant grains abrading and embedding themselves into the skin is described as tattooing; burning of the skin and its hairs is described as singeing.

If the muzzle of the weapon is in contact with the skin at the time of discharge then it may leave a muzzle impression, which is an imprint bruise-abrasion surrounding the skin defect of the gunshot wound. The mechanism producing a muzzle impression is the passage of discharge gasses into the wound track with resultant billowing out of the skin against the muzzle. In these circumstances the stretched skin may lacerate leaving an irregular stellate entry wound. Gunshot wounds through the skull, ribs, sternum and pelvic bones may leave coned defects with a smaller hole on the entry side of the bony plate and a larger bevelled defect on the other side.

The appearance of a gunshot entry wound will depend upon the range of fire, and determining the range of fire from the wound appearance is of forensic importance. If the shot is from a distance then only the bullet will strike the victim leaving a skin defect, typically with an abraded margin or with micro-tears in a high velocity wound, and possibly a grey ring of bullet wipe. At close range, which is within a muzzle to target distance of about arms length or 3 feet (1 metre), the defining characteristic is propellant tattooing on the skin surface. This feature may be blocked by head hair or overlying clothing, although the propellant grains commonly penetrate 1 or 2 layers of thin clothing to embed in the skin beneath. Individual grains of gunpowder should be retained for laboratory analysis since they may indicate the type of ammunition and therefore the type of weapon used.

In contact wounds, when the muzzle of the weapon is in contact with the body or only a few centimetres from the body, all of the elements which exit the muzzle on discharge produce their effects on the wound. The edges of the contact wound are seared by gasses and blackened by soot and propellant baked into the skin, and this may form a broader seared blackened zone around the skin defect. Soot may be deposited in a wide zone around the wound. This soot deposit can be easily wiped off if the wound is washed. Washing blood from the wound carefully using a spray of hot water without scrubbing, should wash off the blood but leave behind any soot deposit.

In suicides, the hand steadying the muzzle of the weapon against the skin may show soot deposition on the radial margin of the forefinger and the adjacent surface of the thumb. If the weapon is a revolver then soot deposits may arise also from the cylinder gap (the gap between the revolving cylinder holding the ammunition and the muzzle) when products of discharge emerge in a fan-like pattern at right angles to the long axis of the barrel. Soot, propellant grains, gunshot residues from the primer and gasses including carbon monoxide, as well as fragments of overlying clothing, are forced into and along the wound track.

Shotguns

Shotguns have a smooth bore and are designed to fire multiple pellets, although they can fire a single slug. The calibre or internal diameter of the bore of a shotgun is measured by an archaic system of gauge. The gauge of a shotgun is the number of lead balls of the given bore diameter required to make up one pound weight. The most popular shotgun gauge is 12, which represents a bore diameter of 0,729 inches (18,2 mm). Confusingly a shotgun with a bore diameter of 0,410 inches (10,2 mm) is not described by its appropriate gauge but as a «four-ten». At the muzzle end of most shotgun barrels is a partial constriction of the bore known as a choke, whose purpose is to restrict the spread of the fired pellets.

Shotgun ammunition, or shells, differ in some design aspects from cartridges for rifled weapons. The cartridge case has a brass lower part, or head, containing the primer mechanism and overlying propellant, and above this is a tube, traditionally made of compressed paper but nowadays of plastic. This paper or plastic tube is closed at the top to retain the lead shot within. Lead shot comes in a variety of sizes which broadly fall into two categories, namely bird-shot, which is used for birds and small game, and buck-shot which is used for large game. The standard pellet sizes are designated by a numbering system, and an individual cartridge bears the appropriate number to indicate the size of shot within. Bird shot pellets range in size from 0,05 inches to 0,18 inches in diameter. Buckshot pellets range in size from 0,24 inches (number 4) to 0,36 inches in diameter (number 000). By recovering a representative sample of the lead shot, determining the average weight and comparing this with a standard table, the size of the lead shot, a characteristic of the ammunition used, can be determined.

Ammunition in the form of single lead slugs is available and an improvised «home made slug» can be produced by partially incising the shotgun shells circumferentially just above the brass head so that this line of weakness breaks on firing and the lead shot exits the barrel still within the otherwise intact cartridge. Although single round lead bullets fired from smooth bore weapons are generally obsolete, they are still used in 0,410 inch muskets for riot control in India.

Within the shotgun shell the propellant and pellets are separated by plastic or cardboard wadding. When the shotgun is fired, these wads exit the muzzle and travel a short distance, sometimes up to 20 feet. At ranges up to about 8 feet the wadding may enter the wound but at greater ranges it impacts the skin adjacent to the wound to leave a wad abrasion. Modern plastic wadding forms a tube around the pellets which opens like the petals of a flower on exiting the muzzle and leaves distinctive petal mark abrasions around the shotgun wound or on the adjacent skin surface. The diameter of cardboard or plastic wadding recovered from a wound or crime scene gives an indication of the gauge of the shotgun used, and may give some indication of the ammunition manufacturer.

Shotgun wounds

In shotgun injuries the pellets rarely exit and therefore the wounding effect is the full wounding potential. Within increasing range, fewer pellets strike the target due to pellet dispersion, and the velocity of the pellets falls off rapidly due to poor pellet aerodynamics. As a result, the range of fire has a critical effect on injury severity in shotgun shootings. At ranges of 20 to 50 metres and beyond there is a uniform peppering of shot which is rarely lethal.

Shotgun wounds of entry have a skin defect and abraded margin and, depending upon range of fire, may show soot smudging, propellant tattooing, singeing from hot gasses and a muzzle impression as with gunshot wounds from rifled weapons. Shotgun wounds display other features reflecting the fact that they are produced by multiple pellets rather than a single bullet, and sometimes show wad marks from the cardboard or plastic wadding of the shotgun shell. Contact shotgun wounds to the torso appear externally innocuous with a circular defect approximating the weapon bore. A muzzle imprint is commonly seen and if there is loose contact or clothing between the muzzle and the skin then there is surrounding soot smudging. The large amount of carbon monoxide forced into the wound produces a faint cherry-red discolouration of the underlying muscle due to the formation of carboxyhaemaglobin.

Contact shotgun wounds to the head are extremely mutilating due to the large amount of energy dissipated by the shot and the large volume of gas which enters the wound. The great majority of wads and pellets will exit and typically the cerebral hemispheres are eviscerated. The shotgun entrance wound to the head is easily identified by the large amount of surrounding soot, the seared and blackened wound edges and the radiating large lacerations. By contrast the exit wound is often not identifiable because of the disruption of the head and fragmentation of the skull. When the entrance wound is intra- oral there are stretch lacerations in the naso-labial folds and at the margins of the mouth.

Propellant tattooing indicating a close range shotgun wound begins at a muzzle to body distance of about 1 to 2 cm and disappears at a range beyond about 1 metre. Beyond this range the appearance of the wound reflects the spread of pellets and any impact of wadding. As the range increases the round skin defect develops scalloped margins (hence its description as a «rat hole» wound in the USA), and then a few adjacent satellite pellet holes before a definite cuff of satellite pellets appears at a range of about 2 metres. At greater ranges there is considerable variation in the pellet pattern depending upon the ammunition used, the degree of choke of the shotgun and the actual weapon. The pattern of spread of the pellets is documented by photography, and measurements of the diameter of the spread are made ignoring aberrant outlying pellets, so called «fliers».

The peeled back petals of plastic wadding may produce a patterned abrasion, in the shape of a Maltese cross, around the wound with the wadding entering the wound at ranges up to 2 or 3 metres. At greater ranges the trajectory of the wadding deviates from that of the pellets so that it impacts the skin adjacent to the wound leaving an abrasion. Precise estimates of range of fire in shotgun injuries require a comparison of the wound pattern on the skin and clothing with test patterns made by firing the same shotgun with the same ammunition.

If the pellets strike an intermediate target, such as a pane of glass or wire door-screen, before striking the victim then this will increase the dispersion of pellets making range assessment liable to error. The dispersion of pellets within a body as indicated by X-ray is always greater than that seen on the skin surface as a result of the «billiard ball effect» produced by the pellets bouncing off each other as they bunch up on entering the body. Shotgun exit wounds are usual where the entry is to the head or neck but exceptional where the entry is to the trunk, unless the trajectory is tangential through the body or there is a contact wound in a thin person or with buckshot ammunition.

X-rays

X-rays of gunshot victims, both living and dead, provide valuable information and X-rays should always be performed in autopsies on gunshot fatalities if the facilities are available. X-rays identify any projectiles present and their location, or may identify projectile fragments if the bullet has exited. The path of the projectile through the body can be determined and often the type of ammunition or weapon used. Projectiles do not necessarily follow a straight line path in the body but may ricochet off bones, most commonly the inner table of the skull, or enter the cardiovascular system and embolise. If the bullet is intact then the exact calibre cannot be determined due to the magnification effects of the X-ray.

The presence of pellets generally indicates the use of a shotgun, although novel shot- containing ammunition is available for some rifled weapons. A «lead snow-storm» comprising scores of lead fragments along the wound track as a result of fragmentation of a partial jacketed bullet is a diagnostic but not ubiquitous feature of high velocity rifle hunting ammunition. Military bullets are specifically designed not to break up in this way within the body but they may do so under some circumstances. With partial metal jacketed bullets, the jacket and core can separate in the body and the lead core exit leaving behind the jacket, with its important rifling marks. Large shotgun slugs are distinctive with designs specific to the manufacturer. The components of some projectiles, such as aluminium jackets, plastic tips and plastic shot shell wadding, are poorly radiopaque and difficult to identify on routine X-rays.

Bullet embolisation is uncommon but may occur within the arterial or the venous system. Arterial embolisation occurs when the bullet enters the heart or aorta and typically embolises to the legs. Venous embolisation occurs when the bullet enters a large vein such as the vena cava or a cerebral sinus and then embolises to the right heart or pulmonary arteries. Small calibre, low velocity projectiles such as a 22 calibre bullet or birdshot pellets are most likely to embolise.

Ballistics

Rifles and handguns, but not shotguns, have rifled barrels. The rifling comprises spiral grooves cut the length of the interior or bore of the barrel. The raised metal between the groves is the lands, and the calibre of a weapon is measured from land to land. When a rifled weapon is discharged, the rifling produces marks on the bullet as the bullet passes down the barrel. Some rifling marks have class characteristics indicative of the make and model of the firearm. Other markings have individual characteristics which reflect imperfections peculiar to a particular firearm, and may allow the identification of that specific firearm.

The class characteristics of bullet rifling marks include:

- The number of lands and grooves, which are usually between 4 and 6 but range from 2 to 22;

- The diameter of the lands and the grooves;
- The width of the lands and the grooves;
- The depth of the grooves;
- The degree of twist of the rifling, that is the number of inches or centimetres of bore required for one complete rifling spiral;
- The direction of the rifling twist, which is commonly a right or clockwise twist and less commonly a left or counter-clockwise twist.

The individual characteristics of bullet rifling marks include:

- Imperfections of the grooves, which are most readily seen in lead bullets as a result of the soft lead being pressed into the grooves;
- Imperfections of the lands, which are most readily seen in jacketed bullets because the copper alloy jacketing is harder.

In a shooting in which a jacketed bullet has been used and the jacket has separated from the lead core, then it is the jacket rather than the lead core which must be recovered for ballistic comparison. Linking a fired bullet to the suspect firearm requires the comparison of the recovered bullet with a test bullet. The test bullet is obtained by taking similar ammunition and firing it from the suspect weapon. The rifling markings on the test bullet and the crime bullet are then compared using a comparison microscope. This type of microscope has a split field in which the crime bullet and test bullet rifling markings can be viewed simultaneously and the bullets rotated to align the markings. If the usually 26 to 28 inch (66 to 71 cm) long barrel of a shotgun is sawn off to aid concealment of the weapon for criminal purposes, then the rough edges around the newly fashioned muzzle may leave marks with individual characteristics on the exiting plastic wadding, permitting a ballistics comparison similar to rifling marks.

Cartridge cases also have markings with class and individual characteristics which include:

- The type of breach block marking;
- The size, shape and location of extractor marks;
- The size, shape and location of ejector marks;
- The size, shape and location of firing pin marks.

Fingerprints are rarely recovered from firearms but may be obtained from cartridge cases, which are necessarily gripped between the fingers when the weapon is loaded. Cartridges too small to retain useful fingerprints may have recoverable DNA.

TESTING FOR GUNSHOT DISCHARGE RESIDUES

The primer, which explodes on compression and ignites the propellant, is commonly a mixture of lead styphnate, barium nitrate, and antimony sulphide. The metals lead, barium and antimony are vaporised and exit the muzzle, to condense and settle on surrounding objects including the hands of the shooter. Bullet metals also may be included in these gunshot residue particles.

Cotton tipped swabs moistened with either 10 % hydrochloric acid or 5 % nitric acid are used to recover residues from the hands of suspects. When taking the hand- wipings plastic gloves must be worn. The 4 areas swabbed are the backs of both hands and the palms of both hands, and each area is swabbed twice. On the back of the hand, the radial aspect of the forefinger, the dorsal aspect of the thumb and the skin web between is swabbed. On the palm of the hand the palmar aspect of the forefinger and thumb together with the base of the forefinger and thumb and the skin web between is swabbed, or alternatively the entire palm is swabbed. The swabs are then analyzed for traces of the elements lead, barium and antimony which characterize gunshot discharge residues.

Flameless atomic absorption spectrometry (FAAS) is used to detect all 3 metals, and neutron activation analysis (NAA) can be used to detect antimony and barium only. Both NAA and FAAS are quantitative elemental analytical methods which do not distinguish the source of the metals, so that false positives are theoretically possible. Barium is a common element found in soil and lead is a common environmental pollutant from motor vehicle exhaust fumes. An alternative analytical method is to recover the residues using adhesive material and examine them using scanning electron microscopy (SEM) and elemental dispersive X-ray analysis (EDAX), which can identify gunshot residue by its characteristic appearance as well as elemental content.

In practice gunshot residues are detectable in about 90 % of people who have fired handguns, but only about 50 % of people who have fired rifles or shotguns. It is not only the presence and quantity of the elements, but also their distribution on the hands which is evaluated. Residues on the palms of the hands only may indicate a defence gesture, or alternatively the handling of a previously fired weapon. In cases of suicide, residues are often detected on the non-firing hand used to steady the muzzle against the body. Recently lead-free primer has been developed because of health and safety concerns about the accumulation of lead in the atmosphere in firing ranges. This novel primer contains titanium and zinc. Firing different ammunition in succession from the same weapon produces residues with a mixture of elements from the different primers and bullets used.

Gunpowder

Gunpowder is a generic term for any one of several low-explosive mixtures used as propellant in guns. The 1st gunpowder was black powder which is a mixture of saltpetre (potassium nitrate), sulphur, and charcoal (carbon) in the approximate proportions 75:15:10. The saltpetre was originally extracted from compost piles and animal wastes.

The consensus is that black powder originated in China, and was being used in fireworks and signals by the 10th Century. By about 1300 the Arabs had developed the 1st real gun, a bamboo tube reinforced with iron. Whether knowledge of black powder entered Europe by contact with the Mongols or the Arabs is unclear but the English medieval scholar Roger Bacon recorded its recipe in 1242 in a coded form of Latin.

Black powder is relatively insensitive to shock and friction and has to be ignited by flame or heat. Its burning is a surface phenomenon so that the fineness of the granulation influences the explosive effect. The product of burning is 40 % gaseous and 60 % solid, the latter accounting for the large amount of whitish smoke produced. Beginning in the middle of the 19th century black powder was supplanted by nitrocellulose-based propellants. Nitrocellulose is manufactured by nitrating cellulose fibres from cotton or wood pulp using nitric and sulphuric acids. It is inherently unstable and burns by decomposing rapidly, forming hot gasses. By contrast with black powder, it produces almost all gas upon combustion, from which it obtained the name smokeless powder.

Today most forms of gunpowder are either single-based, that is nitrocellulose alone, or double-based, comprising a combination of nitrocellulose and nitroglycerine. Glyceryltrinitrate, or nitroglycerine, is a detonating or high explosive characterised by extremely rapid decomposition and the development of high pressure. It is an important ingredient of most forms of dynamite but is employed also in medicine as a vasodilator in the treatment of angina pectoris. Nitroglycerine is 18,5 % nitrogen and contains sufficient oxygen to oxidise the carbon and hydrogen atoms in its molecular structure while at the same time liberating nitrogen. Nitroglycerine was first prepared in 1846 by the Italian chemist Ascanio Sobrero, but it was the Swedish scientist Alfred Nobel who invented dynamite in 1867. Alfred Nobel died in 1896 leaving the bulk of his fortune in trust to establish the Nobel Prizes.

8

Mitochondrial DNA Examination of Cold Case Crime Scene Hairs

INTRODUCTION

Many cold cases have been re-opened in hopes that DNA profiling of evidentiary material may strengthen a case against an existing but weak suspect or identify new leads and new suspects. "Cold hits" are made when nuclear DNA (STR) profiles of semen, blood, or saliva crime scene samples are linked to convicted felon DNA profiles that are stored in the national DNA database (CODIS). Even when actual evidence-to-felon hits are not obtained, inter-case hits between crime scene samples can be useful in investigating serial crimes, where a single unknown perpetrator has left biological material at multiple scenes. No other single technology has been more valuable to cold case investigators than nuclear DNA profiling.

A lesser-known form of DNA testing, however, also is being used for cold case investigation. In the 1990s, mitochondrial DNA (mtDNA) analysis was introduced for samples that had traditionally been unsatisfactory for STR profiling. The earliest use of mtDNA analysis was for the identification of human skeletal remains which contained insufficient or degraded nuclear DNA, but sufficient mitochondrial DNA to aid in matching an individual to his or her maternal relatives. Since 1993, the Armed Forces DNA Identification Laboratory in Rockville, Maryland, has been using mitochondrial DNA to return the skeletal remains of military dead to their families. Identification of missing persons is also aided by this technology. A cutting from a skeletal sample is analyzed, and in the event of a match, the mitochondrial DNA profile of the bone will be the same as the profile of a mother, sibling, or other maternal relative of the individual.

DNA is difficult to recover from very small or environmentally challenged samples. It is degraded or destroyed by heat, moisture, acidity, and fungal or bacterial overgrowth, and its preservation is aided by cold, arid conditions. While nuclear DNA is present in only two copies per cell, the small circular mitochondrial DNA molecule (Figure 1) is present in hundreds to thousands of copies per cell and is therefore a naturally abundant DNA molecule.

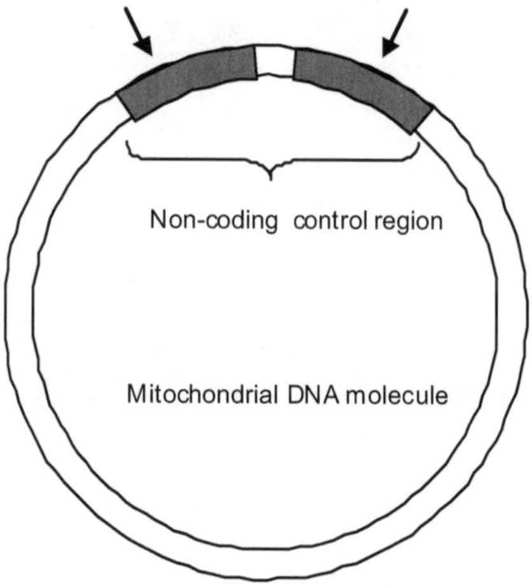

Fig. Schematic diagram of the circular mitochondrial DNA molecule, which contains double stranded DNA that is 16,569 DNA bases long. Forensic analysis examines the order of the DNA bases in two hypervariable regions that characterizes a maternal lineage.

Though all DNA breaks down over time, this natural abundance means that usually enough copies of mtDNA remain for capture and analysis by the forensic laboratory when nuclear DNA is gone. During the 1990s, forensic scientists learned that while naturally shed human hair roots do not contain sufficient nuclear DNA for routine STR typing, they contain abundant mitochondrial DNA. An additional valuable discovery was that shed hair fragments of all kinds with no root are just as useful for this type of testing as hairs with naturally shed roots (some hairs with large, fresh, plucked roots and/or follicular sheath material can be successfully tested for STRs) (Figure 2). Today, the ability to perform mtDNA analysis on virtually any hair is a bonus technique in the investigation of criminal cases. In addition to head and pubic hairs, body hairs, eyelashes, nose hair, and eyebrow hair are excellent candidates for mitochondrial DNA analysis.

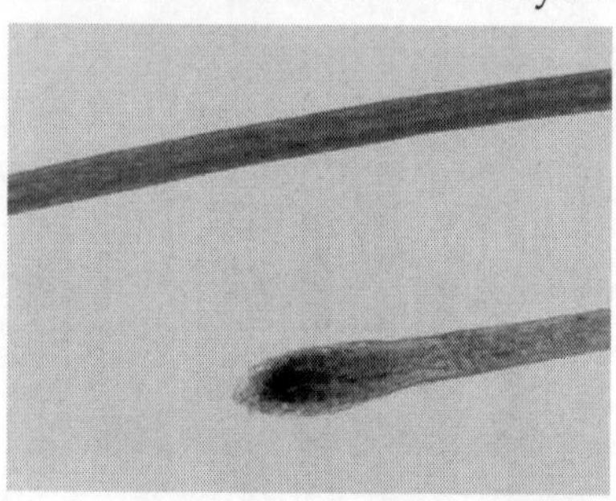

Fig. Photograph of hair shaft (top) and naturally shed hair root (bottom). These are the kinds of hair samples that are analyzed for mitochondrial DNA. [Credit to Max Houck/West Virginia University]

The first FBI trial that used mitochondrial DNA aided in the conviction of a suspect charged with the 1996 rape and murder of a young child in Tennessee. In this case, a pubic hair found in the victim's throat could not be excluded as having come from defendant William Ware. Since that case, hundreds of cases have been examined using mitochondrial DNA analysis, and dozens have successfully been tried in the courtroom. A sizeable body of peer-reviewed scientific literature on the forensic use of mitochondrial DNA is available, and courtroom admissibility hearings, while still underway in some jurisdictions, have uniformly allowed its courtroom introduction throughout the United States.

DNA ANALYSIS

The majority of cells making up the human body are diploid cells carrying identical DNA, with the exception of haploid gametes (egg and sperm) and red blood cells (which have no nucleus). Several types of biological evidence are commonly used in forensic science for the purpose of DNA analysis, including blood, saliva, semen, skin, urine and hair, though some are more useful than others. The use of biological evidence in DNA and genetic analysis varies, with areas of study including blood typing, gender determination based on chromosome analysis (karyotyping), DNA profiling and, more recently, forensic DNA phenotyping. Since the advent of DNA profiling in the 1980s, it has been successfully utilised in criminal cases, disaster victim identification and paternity testing to name a few. However despite their merits, DNA fingerprints are not ideally used as the sole piece of evidence in a case, and in certain countries, such as the United Kingdom, DNA fingerprints must be presented in conjunction with other evidence.

DNA STRUCTURE AND FUNCTION

It is vital to understand the structure and function of DNA and how this relates to DNA analysis in forensic science. DNA, deoxyribonucleic acid, is a molecule arranged into a double-helix, its structure first described by James Watson and Francis Crick in 1953. It is composed of nucleotide trisphosphate molecules, referred to as the 'building blocks' of DNA. These molecules consist of a trisphosphate group, a deoxyribose sugar and one of four nitrogenous bases. The four bases involved in a DNA molecule are adenine and guanine (purines) and thymine and cytosine (pyrimidines). These bases bond to the deoxyribose sugar and one of the other bases to form base pairs, with adenine and thymine bonding through two hydrogen bonds, and guanine and cytosine bonding with three hydrogen bonds.

DNA is essentially the molecule that holds all genetic information and 'instructions' for an organism. The human genome is composed of over 3 billion base pairs of information organised into 23 chromosomes. Genes are the regions of DNA that encode and regulate protein synthesis, though this involves just 1.5% of the entire genome. A significant amount of the human genome, approximately 75%, consists of extragenic DNA, which contains regions that do not actually contain known gene sequences. About 50% of extragenic DNA is made up of something called repetitive DNA, which is of particular use in forensic DNA analysis.

Repetitive DNA is further sub-divided into tandem repeats (including satellite DNA, microsatellites and minisatellites) and interspersed repeats (SINE, LINE, LTR and Transposon). Tandem repeat DNA and the variation between them (polymorphisms) is the focus of many DNA profiling techniques. It is due to the number and location of these polymorphisms that every individual has unique DNA which produces a distinctive band pattern when analysed. It is through the extensive study of the genome that DNA fingerprinting has been produced as a useful and reliable technique in forensic science.

Sources of DNA Evidence & DNA Extraction

In terms of forensic DNA analysis, there is a variety of possible sources of DNA evidence. The more useful sources include blood, semen, vaginal fluid, nasal secretions and hair with roots. It is theoretically possible to obtain DNA from evidence such as urine, faeces and dead skin cells, though this is often classed as a poor source due to the lack of intact cells and high levels of contaminants preventing successful analysis. Such samples will be collected depending on the type of sample (see crime scenes page for more details of evidence collection and preservation).

Prior to analysis, it will be necessary to extract DNA from the sample. *This is generally achieved through the following simplified steps*:

1. The sample cells are lysed (broken down) in a buffer solution.
2. Denatured proteins and fats are pelleted through centrifugation.
3. The cleared lysate is then passed through a column, often containing a positively charged medium that binds to the DNA.
4. Contaminating proteins, fats and salts are then removed through several washes.
5. The DNA is recovered in a buffer solution/water.

The amount of DNA is often then quantified using spectrophotometric techniques. Various methods of extraction have been devised for different types of sample.

DNA samples for comparison are generally collected from suspects using buccal swabs, in which a sterile swab is scraped along the inside of the cheek to collect epithelial cells to use in producing a DNA fingerprint. Sir Alec Jeffreys is often described as the father of DNA fingerprinting.

Single Nucleotide Polymorphisms

The use of DNA analysis in forensic science is based on a variety of techniques focusing on polymorphisms, which essentially refers to variation in sequences. Different sequences are studied in different techniques, including single nucleotide polymorphisms, minisatellites (variable number tandem repeats), microsatellites (short tandem repeats) and mitochondrial DNA, each different with regards to length and repetition.

Single Nucleotide Polymorphisms (SNPs) are the simplest and most common type of genetic variation, composing around 90% of genetic variation in humans. They occur during meiosis when DNA is replicated, with each SNP representing a difference in a single nucleotide. For example, a SNP may replace cytosine with a thymine in a stretch of DNA, and this will not change the length of the DNA. There are about 10 million SNPs in the human genome, with one found at every 100-300 base pair. These can act as biological markers.

In the analysis of SNPs, which can consist of up the four alleles, the specific base present in the SNP must be established. This is achieved using DNA sequencing techniques. However DNA sequencing is not generally used in forensic science except in the analysis of mitochondrial DNA.

Dideoxy Method

Also known as the Sanger Method, this form of DNA sequencing was developed by Fred Sanger in 1975. A labelled primer is utilised to initiate the synthesis of DNA. Four dideoxy nucleotides are added and randomly arrest synthesis. Fragments are produced which are subsequently separated using electrophoresis or, in more modern automated systems, capillary electrophoresis. Specialised software is then used to convert the band patterns into a DNA sequence. Initially unstable, hazardous radioactive tags were used, soon replaced by fluorescence dyes.

However the varying mobility of these different coloured dyes meant that the order of bands did not necessarily accurately represent the order of the nucleotide sequence. The dyes would also sometimes behave unpredictably, making them unreliable and not particularly ideal for use in forensic cases. If conditions were unsuitable for the DNA template or the dideoxy nucleotides, the reaction would be unsuccessful. The computer software itself was problematic, with peaks often overlapping, making it difficult to establish the correct order of nucleotides.

Minisatellites or Variable Number Tandem Repeats (VNTRs)

Variable Number Tandem Repeats (VNTRs) are highly polymorphic sequences in the exon region of DNA repeated at various points along the chromosome, 4-40 times. They are about 20-100 base pairs in length, though their specific lengths are not strictly defined. These tandem repeats are inherited from both parents, therefore no one will have the same VNTRs as either of their parents. The number of repeats at the locus affects its position after electrophoresis and the length of the DNA after the chromosome has been cut with a restriction enzyme. Minisatellites were the first type of polymorphism to be used in a criminal investigation in the case of British murderer and rapist Colin Pitchfork.

However VNTRs are not generally used in forensics, as this technique is often a more expensive process and takes longer due to VNTRs having a greater length than, say, STRs.

Microsatellites or Short Tandem Repeats (STR)

Short Tandem Repeats (STRs) are regions of the genome composed of approximately 1-5 bases and repeated up to 17 times. STR markers will either be simple (identical length repeats), compound (two or more adjacent repeats) or complex (several different length repeats). They are found on 22 autosomal chromosomes as well as both X and Y sex chromosomes, though those on the Y chromosome differ less due to lack of recombination. Only a select number of STR markers are used in forensic DNA profiling (10 in the UK and 13 in the US).They express a high degree of polymorphism, making them of particular use to the forensic scientist. The variability in STRs is caused by the inaccuracy of DNA polymerase in copying the region. As STR regions are non-coding, there is no selective pressure against the high mutation rate, resulting in high variation between different people.

Though there have been thousands of short tandem repeats found in the human genome, only a small number are utilised in forensic DNA analysis. STRs used in forensic science tend to be tetra- and penta-nucleotide repeats, as they are both robust, suffer less environmental degradation, and provide a high degree of error free data. STR loci are ideal for use in forensic science for a number of reasons. They represent discrete alleles that are distinguishable from one another, they show a great power of discrimination, only a small amount of sample is required due to the short length of STRs, PCR amplification is robust and multiple PCR can be used, and there are low levels of artefact formation during amplification. An early use of microsatellites is in the identification of Auschwitz camp doctorJosef Mengele.

Restriction Fragment Length Polymorphisms (RFLPs)

Restriction Fragment Length Polymorphisms (RFLPs) were used in the first technique developed to analyse variable lengths of DNA fragments produced through DNA digestion. It exploits variations in DNA sequences due to the differing locations of restriction enzyme sites. The method uses restriction endonucleases to 'digest' the DNA by cutting it at specific sequence patterns. The resulting restriction fragments are then separated using gel electrophoresis and transferred to a membrane using the Southern Blot technique. After the separated DNA fragments are transferred, probe hybridisation is used to detect the fragments.

However DNA analysis with RFLP required relatively large amounts of DNA and degraded samples could not be analysed with accuracy. More effective, faster and cheaper DNA profiling techniques have seen been developed, so RFLP is generally no longer used in forensic science.

Polymerase Chain Reaction (PCR)

The amount of DNA evidence obtained during the investigation of a crime is often very small, thus for successful DNA profiling some form of amplification is ideal. Polymerase Chain Reaction (PCR) is a technique which allows for the exponential amplification of DNA fragments to lengths of approximately 10,000 base pairs. This means that, theoretically, a single copy of a DNA fragment could be amplified to millions of copies in just a few hours. PCR is particularly beneficial in the amplification of minute amounts or degraded samples.

A successful PCR reaction requires a number of vital primary components. Oligonucleotide primers which are complementary to the DNA target and mark the target to be amplified, with two primers being used. The base sequence of one primer binds to one side of the target whilst the other primer binds to the other side of the target, with the DNA between the primers being amplified.

Fluorescent tags are often added to the primers to visualise amplified DNA in electrophoresis. DNA polymerase enzyme allows the DNA strand to be copied by adding nucleotides to the 3′ end of the primers. Other components required include a reaction buffer with MgCl to ensure ideal conditions for the functioning of the DNA polymerase enzyme, deoxyribonucleotides to build the DNA molecule, and template DNA. Modern PCR uses thermostable DNA polymerases. Most commonly used is the Taq polymerase, which has largely replaced the previously used E.coli-derived polymerase.

This was isolated from Thermus aquaticus, which is an organism capable of surviving in temperatures over 70oC. However Taq polymerase lacks the ability to proof read. VENT polymerase is from Thermococcus litoralis, which can survive in temperatures over 100°C.

The PCR cycle consists of three primary steps: denaturation, annealing and extension. The process is generally conducted in a small, plastic centrifuge tube with the temperature carefully controlled using a thermal cycler.

- *Denaturation*: The sample is heated to 94-95oC for about 30 seconds. This separates the double-stranded DNA by breaking hydrogen bonds, allowing primers access.
- *Annealing*: The samples is kept at 50-65oC, depending on the primer sequence, to allow hydrogen bonds to form between the primers and the complementary DNA sequence.
- *Extension*: Also known as the elongation stage. The sample is heated to 72oC for a duration depending on the length of the DNA strand to be amplified and the speed of the polymerase enzyme (Taq polymerase) which builds up the strand. Deoxynucleotide triphosphates are added to the 3′ end of the primer.

Each PCR cycle can take only 5 minutes. This procedure can then be repeated as necessary until the original sequence has been amplified a sufficient amount of time, with the amount being doubled with each cycle. Following PCR, the products are separated using electrophoresis. Unfortunately PCR is not suitable in the analysis of longer strands of DNA, and so cannot be used with earlier techniques such as RFLP. It must be taken into consideration that certain compounds can inhibit PCR reactions, often substances associated with the stages of extracting and purifying the DNA. Such substances include proteinase K (which degrades the polymerase enzyme), ionic detergents and gel loading dyes. Similarly, certain substances present in blood can inhibit PCR, such as haemoglobin and heparin.

Various alterations have been made to improve the PCR method. Multiplex Polymerase Chain Reaction involves the amplification of numerous DNA sequences in a single reaction through the use of primers that produce non-overlapping allele sizes, allowing numerous regions of a sample to be tested simultaneously.

PCR Errors

Various factors can contribute to errors and inaccuracies in data produced by the polymerase chain reaction technique. PCR is often carried out using DNA polymerases such as Taq DNA polymerase, which does not have the ability to 'proof read', resulting in errors in amplification. The greater the amplification, the more likely it is that such errors will occur. Mispriming is also a potential problem, with products being formed from non-target sites. Excessive primer dimers may be formed, which are by-products of PCR produced when one primer is annealed to another causing primer extension. This may all result in unexpected variability in PCR success across a series of samples or previously successful conditions failing.

Electrophoresis

As mentioned, during DNA analysis the individual fragments of DNA can be separated using electrophoresis to produce the distinct 'DNA fingerprint'. Electrophoresis is essentially a method of separating molecules by their size through the application of an electric field, causing molecules to migrate at a rate and distance dependent on their size. In gel electrophoresis, a porous gel matrix is used, often consisting of agarose gel for simple work or polyacrylamide gel for more specific procedures. The gel is often floating in a buffer solution to ensure the pH level is maintained and the applied electric current is conducted. Samples to be analysed are placed in small wells at the top of the gel using pipettes.

A control sample and a standard/marker sample will often be run simultaneously. As the electric current is applied, the negatively charged DNA fragments begin moving through the gel towards the positively charged anode. The gel essentially acts as a type of molecular sieve, allowing smaller molecules to travel faster than larger fragments. Following electrophoresis, it may be necessary to visualise these bands using radioactive or fluorescent probes or dyes. Electrophoresis not only separates DNA but also allows for the fragments to be measured, often expressed in base pairs. Measuring the length of these fragments can ultimately allow the number of repeats to be determined and thus the genotype at that locus.

Earlier techniques used flat-bed gel electrophoresis, though the faster and automated capillary electrophoresis is now used more often. In this technique, abbreviated to CE, the gel is held in a fine capillary tube through which the fluorescently-labelled DNA passes through much as in gel electrophoresis, often with an added DNA size standard. Positioned along the capillary is a laser beam which causes the DNA to fluoresce as it passes. This can then be detected and fed directly to a computer system in the form of an electropherogram. Unfortunately capillary electrophoresis is not able to separate more than one sample at a time, though a genetic analyser can be used to separate a series of samples one after the other.

Blotting Techniques

Following gel electrophoresis, probes are generally used to detect specific molecules. However because probes cannot be directly applied to the gel, blotting methods were initially utilised. There are a number of common blotting techniques: Southern blot, Western Blot, Northern Blot, Eastern Blot and Southwestern Blot.

- *Southern Blot*: Used in DNA analysis. DNA is extracted, separated with electrophoresis and transferred to the membrane. Labelled probes are added, which hybridise to specific sequences and identify them.
- *Western Blot*: Used to detect proteins. SDS-PAGE used to separate proteins, which are transferred to membrane. Specific antibodies are then added, followed by a substrate to visualise bands.
- *Northern Blot*: Used to study gene expression. Similar to the Western blot, except RNA is being analysed.
- *Eastern Blot*: Used to study proteins. Considered to be an extension of the Western blot.
- *Southwestern Blot*: Combines features of the Southern and Western blot. Used for the rapid characterisation of DNA binding proteins and their sites.

Low Copy Number (LCN) DNA Analysis

Low Copy Number DNA Analysis, referred to as LCN, is a technique developed by the UK's Forensic Science Service in an attempt to increase the sensitivity of DNA profiling methods. Samples containing small amounts for badly degraded DNA often leads to problems such as poor quality fingerprints or even completely negative results. This technique reduced these issues.

Developed in 1999,LCN is essentially an extension of the Second Generation Multiplex Plus (SGM+) technique, and is generally used when the previous technique has failed or has produced weak results. Improved sensitivity is achieved through an increased number of PCR cycles, with standard techniques generally using 28 cycles but LCN using 34. This could ultimately allow for DNA profiles to be successfully obtained from minute amounts of sample and even from single cells.

However the increased sensitivity of this DNA profiling technique brings about amplified concerns over issues of ease of contamination and amplification of these contaminants, mixed profiles being produced and wrongful accusations. With techniques such as LCN, it is now more important than ever that investigators wear suitable protective clothing and follow strict anti-contamination procedures, and controls are used in analyses.

In 2007, LCN came under great scrutiny through the case of R vs Hoey, which led to the use of this technique being temporarily banned until a thorough review could be conducted. Sean Hoey was tried for involvement in the Omagh bombing in 1998, charged with numerous counts of murder in addition to terrorism and explosives charges. The DNA evidence used against him was based on the relatively new LCN technique which, at the time, was only being used in British courts. However due to the lack of data and known error rates regarding the technique, serious concerns were raised. Hoey was subsequently found innocent.

Single Cell DNA Fingerprinting

Closely linked to LCN analysis is single cell DNA profiling. Dr Ian Findlay and his colleagues at the Australian Genome Research Facility first reported the successful development of a DNA fingerprint from a single cell in 1997. 226 buccal cells from four individuals were isolated using micromanipulation, amplified using the UK Forensic Science Service's Second Generation Multiplex (SGM) assay to increase sensitivity. Six STR markers and amelogenin for sex-typing were amplified. The results from single-cell analysis were compared with known DNA profiles. Whereas a full and correct profile was only obtained in 50% of the single-cell tests, 91% of tests exhibited some form of result.

Previous DNA fingerprinting methods often required hundreds of cells in order to obtain a profile. The single cell is obtained by swabbing the material and identifying the cell to be analysed using microscopy prior to analysis. This technique is particularly fast, taking a matter of hours, and has a specificity of about 1 in ten billion.

Single-cell DNA profiling is particularly useful in rape cases, as DNA in sperm cells is highly conserved due to it being so compacted in the protein head. There is also potential for the technique in use in documents. Human DNA can be placed in documents such as Government bonds, wills and security documents, to track their flow. However the main issue with this particular use is that close relatives may handle the documents, particularly when dealing with documents such as wills, and so the technique may not be appropriate. Single-cell DNA analysis is also ideal in IVF procedures, in which single cells can be analysed for genetic defects.

However there are some major concerns with this method. As an increased number of PCR cycles are required to amplify the DNA, this brings about problems of allele drop-in, where additional alleles are added to the sample. Allele drop-out is a similar problem with increasing amplification samples. When starting with a single cell or small amount of sample, any contaminants already present in the sample will be amplified. Similarly, any PCR inhibitors present will be more concentrated, causing amplification problems. One of the primary concerns is the ease of contamination by cells from other individuals. This could result in samples being contaminated and rendered useless or, worse still in the case of forensics, innocent individuals being wrongly accused. More work is required to validate the technique, particularly to use it in forensic science. The work in single-cell DNA analysis led to the Forensic Science Service in the UK developing low-copy number DNA analysis.

Mitochondrial DNA (mtDNA)

Mitochondrial DNA is a circular molecule of DNA 16,569 base pairs in size, first referred to as the Anderson sequence, obtained from the mitochondrion organelle found within cells. This organelle is involved in the production of cell energy. There can be anywhere between 100 and 1000 mitochondria within a cell, each one containing numerous copies of the mitochondrial genome. This sequence is entirely functional and highly conserved, so there is very little variation between individuals.

However there is a 1000 base pair long non-coding D-loop, known as the control region, which contains two hypervariable regions referred to as HV1 and HV2. The variations within these regions are generally single nucleotide polymorphisms (SNPs). SNPs do not alter the length of the mtDNA, and it is these regions that are focused on in the forensic analysis of mitochondrial DNA. Mitochondrial DNA is often subject to a relatively high rate of mutation due to its lack of DNA reparation, causing variation between individuals. The variation within this small portion is itself not especially significant, with HV1 and HV2 differing by 1-3% between non-related individuals.

In the analysis of mitochondrial DNA, the DNA is extracted and the HV1 and HV2 regions amplified using PCR. The base pair sequence of these regions is then established through DNA sequencing (see DNA sequencing section). This is then compared with the Cambridge Reference Sequence and differences noted. Other samples can then also be analysed an comparisons made to establish potential similarities. Mitochondrial DNA is generally used when other methods such as STR analysis have failed. This is often in the case of badly degraded bodies, in cases of disaster or accidents where an individual is too badly damaged to identify, and sometimes in taxonomy to determine species using the cytochrome b gene.

The most significant advantage of the use of mitochondrial DNA is the possibility of analysing even highly degraded samples. If a specimen is severely decomposed to the point that it is not possible to successfully extract a DNA profile using nuclear DNA, it may be possible through mitochondrial DNA. Additionally, only a very small sample size is required.

However the use of mtDNA does have its disadvantages. As mitochondrial DNA is only maternally inherited, this cannot form a full DNA fingerprint of the individual, thus this technique is only beneficial if the DNA profiles of maternal relatives are available, such as the individuals mother or biological siblings. Because of this, mtDNA is significantly less discriminatory than, for example, Short Tandem Repeats. Detecting sequence differences is also relatively time consuming and expensive.

FORENSIC DNA PHENOTYPING

Attempts have been made to utilise DNA analysis in the identification of phenotypic characteristics such as skin colour, hair colour and eye colour in a study known as forensic DNA phenotyping (FDP) or phenotypic profiling. This is generally achieved using single nucleotide polymorphisms (SNPs) rather than STRs. SNPs have a lower mutation rate and so are more likely to become fixed in a population, thus they are often found to be population-specific. It may be possible to estimate ethnic origin based on the presence of rare SNPs or STRs linked to particular population groups, though this theory becomes problematic with individuals of mixed ancestry.

Most work on phenotype SNPs has focused on pigmentation, and SNPs in a number of pigmentation genes have been associated with various human hair, skin and eye colour phenotypes. Advances are already being made in this area of study. The Forensic Science Service has developed an SNP typing assay involving mutations in the human melanocortin 1 receptor gene (MC1R), which is associated with red hair. Specific alleles along this gene are associated with red hair, thus an individual inheriting this allele from each parent results in a high likelihood of that individual having red hair. Establishing that an individual is likely to have red hair is limited in forensic science, as red hair is not particularly common, though it is more common to certain populations.

Some research has been carried out into the genetics of eye colour, namely relating to the OCA2 gene on chromosome 15, which is also involved in the pigmentation of both skin and hair. IrisPlex is a recent test developed which allows for the accurate prediction of blue and brown eye colour. Studies have been conducted examining the frequency of specific short tandem repeat alleles in groups of varying geographic ancestry. This research has resulted in the possibility of likely ancestry being established due to certain alleles being more likely in specific groups. However this particular use of DNA analysis is not infallible and can only be used as an estimation. The Forensic Science Service has developed an ethnic inference test, which provides the likely origin of a DNA sample from a number of groups (white European, Afro-Caribbean, Middle Eastern, South-East Asian and Indian Sub-continental).

Some countries and states are implementing specific legislation relating to the use of phenotypic DNA analysis. Many jurisdiction presently only allow the analysis of non-coding DNA, though the Netherlands currently allows forensic phenotyping under certain regulations. The phenotypic use of SNPs have also been used in non-forensic analysis, such as in determining the ethnicity, hair and skin colour of ancient remains.

However there are a number of problems and concerns with this approach. It is unlikely that a few chosen SNPs will provide a foolproof 'picture' of the sample's source due to the complexity of multigenic traits as well as external factors such as environment and aging. Even if the technique was perfected for use in forensic science, phenotypes such as hair or eye colour can easily be masked through dyes and coloured contact lenses, limiting its forensic use. There are also privacy concerns relating to the possible traits determined, though it has conversely been argued that visible traits such as hair and eye colour cannot be considered private. Though should further advances be made to the extent that genes were located for traits such as aggression and predispositions to certain diseases, more serious concerns would be raised over the sensitivity of such information.

Researchers anticipate that further advances could allow for additional details to be ascertained, such as the likelihood that the individual smokes, along with the possibility for genes for the likes of handedness, aggression and homosexuality.

Y Chromosome Analysis

Much of modern DNA profiling is based on the analysis of short tandem repeats found on autosomes. However one particular branch of DNA analysis focuses on the amelogenin marker, the only marker on the sex chromosome, useful in the analysis of the Y chromosome. The Y chromosome, generally found only in males, is a small chromosome which, unlike other genes, is only altered through the infrequent occurrence of mutation. However similar to mitochondrial DNA (which is maternally inherited), the combination of alleles in this instance is theoretically identical between father and son, assuming mutation does not occur. Furthermore, Y chromosome analysis discrimination is comparatively low. Y chromosome analysis is particularly useful in cases of sexual assault and rape in which mixed DNA profiles may be encountered. Numerous systems have been developed to analyse some of the STRs present on this chromosome, such as Applied Biosystems' Yfiler.

DNA Databases

Numerous countries have produced computerised databases containing DNA profiles to aid in the comparison of DNA fingerprints and the identification of suspects and victims. The first Government DNA database was established in the United Kingdom in April 1995, known as the National DNA Database (NDNAD). As of 2011, there were over 5.5 million profiles of individuals in the system. Similarly, the FBI in the US formed their own DNA database, the Combined DNA Index System (CODIS), in 1994, though it was not implemented in all states until 1998. Staff members involved in the handling and analysis of evidence will often also submit their DNA profiles to the database in the case of accidental contamination. There is the possibility for DNA databases to be shared between countries, however some countries focus on different loci in DNA fingerprinting.

DNA databases are not only used to make direct matches between the DNA fingerprints of one person, but to also conduct familial searching. This involves the search for genetic near-matches between a victim/suspect and a member of their family whose DNA profile is stored. This technique is based on the principle that related individuals are likely to express similarities in their DNA profiles. The first prominent use of familial searching was in the case of British serial killer Joseph Kappen, who was seized after his son's DNA profile was obtained and then linked to him.

The production of DNA databases has allowed for the faster apprehension of suspects through comparing new crime scene samples to those already stored in the database, providing links between criminals and other crimes. It has also been widely used in cold cases, in some instances proving the guilt of an individual decades after they committed the crime. Conversely, wrongly imprisoned individuals have been exonerated through the advent of new DNA analysis techniques and databases. There is also the potential benefit of identifying bodies that have been too badly damaged or decomposed to identify, provided the individual's DNA profile is stored.

However despite the advantages of such databanks, there has been significant controversy relating to the subject. Standard practice in many locations is to take a DNA sample of an individual when arrested, but this raises concerns over whether their DNA should be retained if they are then found innocent. There are also worries over innocent people being identified as matches or partial matches to DNA found at a crime scene even if they were not involved in the crime but had innocently attended the scene at some other point in time. This particular problem is becoming increasingly concerning as DNA fingerprinting techniques become more sensitive. There are also privacy concerns over the availability of sensitive genetic information, such as susceptibility to certain diseases and familial relationships. A more administrative disadvantage of such databases relates to the need for a facility that is both large enough to store such data but also has adequate security, a combination that can prove extremely expensive.

DNA Profile Interpretation

The primary purpose of forensic DNA profiling is to obtain a DNA 'fingerprint' from a biological sample and compare this to profiles obtained from DNA from a crime scene, an individual or profiles stored on a database. The process of modern DNA profiling includes statistically determining the chance that two people will share the same profile by establishing how common a genotype or collection of genotypes is within a population. The more loci studied and the greater heterozygosity of these loci, the smaller the chance two people will share a profile. However such figures can only ever be estimations and do not take certain factors into consideration, such as biological relatives. DNA evidence tends to be presented in terms of a random match probability, rather than definitively stating whether two profiles match or not.

MITOCHONDRIAL DNA IN COLD CASES

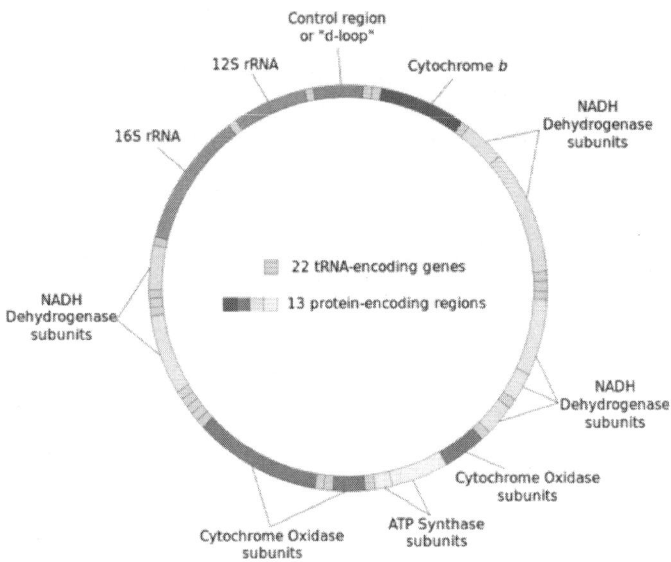

Mitochondrial DNA is found in non nucleated cells, like teeth, hair and bones. It is circular DNA, composed of base pairs. There are only two copies of DNA in nucleated cells, whereas there are hundreds of mtDNA, proving to be a more substantial DNA source than those currently used to find regular DNA. A major benefit of mtDNA is that it degrades very slowly. One major disadvantage to mitochondrial DNA is that all maternal relatives have the same mtDNA profile, meaning that a profile cannot match one individual. However, this is not as big of a problem as it seems, since mtDNA profiles still eliminate over 99% of the population (Melton, 2006) in most cases, and is a good method of ruling out suspects. It is also useful when suspects have been long dead (common in old cases), allowing certain family members to provide samples for mtDNA profiling. DNA profiles are mostly obtained by electrophoresis. There are two main types of electrophoresis, gel and capillary. In both, the DNA is cut in to lots of fragments using a restriction enzyme. Gel electrophoresis involves the injection of the sample in to a well on a gel plate, submerged in a buffer solution. A current is run through the gel, moving the DNA with it through a series of mesh-like layers.

A band appears when lots of fragments of DNA can not fit through the 'mesh', while smallerfragments keep moving. The gel plates are treated to make the bands visible under normal light. This is what is then used to make a comparison between suspect profiles and evidence profiles.

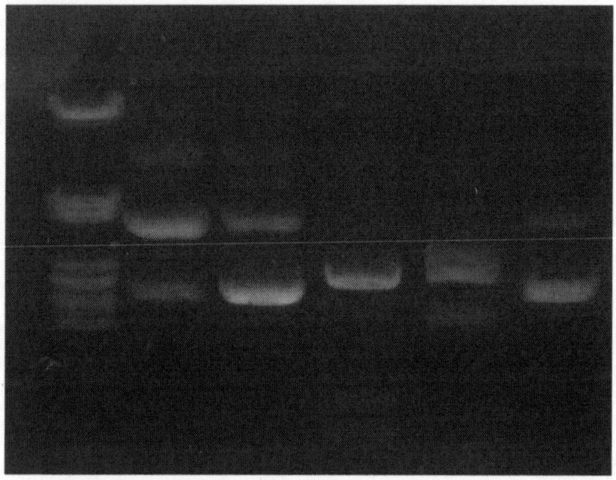

Capillary electrophoresis is similar to gel, but fragments are separated in a capillary, not a gel plate, and separated based on charge, rather than size. Capillary electrophoresis is faster than gel, however only one sample, plus a 'standard' DNA ladder can be run per capillary, compared to the several that gel can handle.

PCR is used to make millions of replicates of DNA,without affecting the original copy. In some cases only a few skin cells are needed to start the reaction. Cold case DNA usually requires special treatment, low copy number (LCN). This method detects the same ten regions of DNA, plus a gender marker, that SGM+ does, however the PCR stage uses 34 cycles, to make sure that there is enough DNA to work with efficiently.

Each PCR cycle doubles the amount of DNA. Due to the high risk of contamination when performing LCN, duplicate tests are carried out whenever possible, to ensure that no contamination has occurred. Only if two tests return the same DNA profile can it be said that sample was not contaminated. In 2007 a review, and the suspension, of the use of LCN was carried out by recommendation of a judge. LCN use continued in 2008, after the review found that it was still a very reliable method of obtaining a profile.

USE OF MITOCHONDRIAL DNA IN FORENSICS

As per forensic science is concerned, investigators left no stone unturned, while solving cases. DNA-based technologies have been used to identify victims of a crime. In such methods, the DNA sequence of individuals or samples available from the crime scenes, also known as DNA fingerprint, is scanned to identify the victim. Most popular and effective DNA technologies that are used in forensics include RFLP (restriction fragment length polymorphism), PCR (polymerase chain reaction) and STR (short tandem repeat) analysis.

In case of very old biological samples, which are devoid of nucleated cellular material, mtDNA analysis is used. In recent times, the technique of mtDNA analysis is used to identify victims of cold cases, mass disasters (tsunami and earthquakes), historically important cases (war soldiers) and cases with limited biological material. Another advantage of mtDNA is that it can be isolated from samples like hair shafts, bones and teeth. Thus, investigators can study these samples by extracting mtDNA.

In comparison to extracting nuclear DNA, mtDNA isolation is more easy due to the presence of multiple copies in a cell. In humans, there are two copies of nuclear DNA (nDNA) in a cell; whereas, the mtDNA copies range from 100-10,000 per cell (except in egg and sperm cell). Hence, very small samples can be tested for victim identification. In general, a mitochondrion contains at least 2-10 copies of DNA.

Since mtDNA is inherited from the mother's egg cell, it can be used to establish the maternal lineage. In simpler terms, the mtDNA of both mother and daughter are same. Hence, mtDNA analysis serves as a valuable tool, while investigating the case of a missing person. The mtDNA from the unidentified remains can be compared with the mtDNA profile of a maternal relative (such as brother and sister, mother and her daughter). If the mtDNA profiles of the two individuals match, then they are relatives. Thus, any maternal relative will be able to provide a reference sample.

These are the advantages of using mtDNA as a crime-solving tool in forensic studies. The mtDNA, because of its circular structure and location inside the cell, is more stable and strong than the nuclear DNA. It is shown that mtDNA is protected from degradation, even when exposed to prolonged environmental conditions, which is not the case with nDNA. This is the reason why, mtDNA is present even in older and degraded specimens, whose nDNA is already degraded. The only limiting factor of mtDNA analysis is that there is little discrimination between the mtDNA sequences of the same maternal lineage.

ADVANTAGES AND LIMITATIONS OF MTDNA ANALYSIS

A nuclear DNA match of the 13 core STR loci permits little doubt that a questioned sample has come from a known individual, except when identical twins must be discriminated. However, because mitochondrial DNA is maternally inherited, all a woman's offspring, her siblings, her mother, and other maternal relatives will have the same mitochondrial DNA profile. Mitochondrial DNA, therefore, is not a unique identifier in the way that nuclear DNA is, and the test's conclusion can be only whether or not a known individual is excluded as the donor of the questioned sample. In testimony, a mtDNA forensic examiner will state whether the known individual (and his or her maternal relatives) could or could not have deposited a crime scene hair, but can never state unequivocally that a hair came from a particular person. In fact, because we lose track of our distant maternal relatives, we must assume that an apparently unrelated individual may have the same type we have; this means that exclusionary testing of the victim, victim's partner, victim's non-maternal relatives, and alternative suspects is often necessary when mtDNA analysis is used.

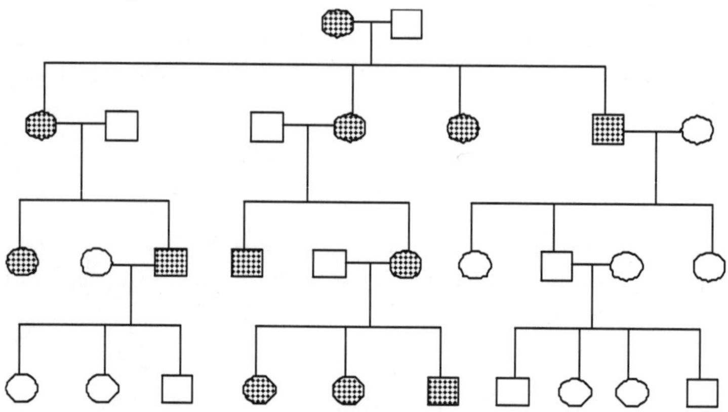

Fig. Pedigree chart showing the maternal inheritance of mitochondrial DNA. Women are circles and men are squares. The individuals with the pattern have the same mtDNA type, and can all be traced maternally to the woman in the top line. Note that men inherit their mother's mtDNA type, but do not pass it on to their children

Because a mtDNA type is found in maternally related individuals, it is unlikely that a felon database, such as the national CODIS system, will ever be developed for mitochondrial DNA cold-hit searching. Because of this, in any given case investigators must develop a list of potential donors to be compared to the analyzed questioned hairs, such as victims and suspects, their relatives, and even elimination samples for crime scene personnel. The probative value of the questioned hair will dictate the range of known samples that should be considered for comparison to the questioned samples.

While non-uniqueness is a limitation, there are thousands of different mitochondrial DNA types, and the relative population frequency of almost any type is low. In most cases, at least 99% of the population will be excluded as contributors and the pool of random individuals who could have contributed the sample is small (less than 1%). In many cases, well over 99.9% of the population may be excluded. The inability to use mtDNA in quite the same way as nuclear DNA highlights its value as a "part of the puzzle", meaning that mtDNA almost always supplements other information in the theory of the crime and that evidence tested by this method would rarely be the only evidence.

An advantage, however, of mitochondrial DNA use is that if a victim is missing (no-body homicide), a single maternal relative may donate the mtDNA reference sample to compare to suspected crime scene victim hairs. When a suspect is long deceased, missing, or unavailable for other reasons, his or her maternal relative may provide the mtDNA reference sample for comparison to crime scene samples. Alternatively, exhumed skeletal remains such as teeth, femurs, or ribs from individuals who have no known living maternal relatives, while usually unsuitable for nuclear DNA analysis, are almost always satisfactory for mtDNA use as known reference samples, although collection is more difficult and costly.

CRIME SCENE HAIRS

A recent development in the area of post-conviction relief is the mtDNA re-examination of shed crime scene hairs previously examined a trace examiner using only a microscope. In a number of older cases, exonerations or new trials have been won when mtDNA analysis of a hair has proven without any doubt that a convicted offender could not have left the hair in question. In these cases, during trial, a trace examiner had testified that the hair "matched" the defendant and the jury weighed this limited testimony heavily. A recent study by the FBI, however, showed that such microscopic evaluations, in their highly experienced trace evidence examiner's hands, result in false positives about 11% of the time when the match was tested using mtDNA analysis.

Similarly, false negatives also occur: that is, hairs that might have shed light on a case were discounted by the examiner because the microscopic evaluation showed no visual "match" between an individual and a crime scene hair. Because the range of physical variation in hair samples from a single individual is vast, it is conceivable that physical differences of color, diameter, and structure between hairs from widely diverse locations of the body could be wrongly interpreted.

The knowledge that false negatives and false positives may occur with hair microscopy provides an opportunity for the cold case investigator. By re-examining old records, laboratory and expert reports, stored evidence, and transcribed testimony from interviews, depositions, hearings, and trials, the investigator may locate previously slide-mounted hairs, loose hairs on clothing or in envelopes or paperfolds, or other crime scene material that was previously discounted on the basis that either "one can't do anything with shed hairs/hairs with no root" or "the trace examiner said there was no match". These sample hairs can open up new avenues for consideration of alternative or weak suspects who may have left them at the crime scene. Similarly, crime scene hairs may help place a victim in probative relevant locations. Because no hair root is necessary for mtDNA analysis, many previously discounted smaller and fragmented hairs have recently become valuable pieces of evidence.

Examples are:

- Hair under a victim's fingernails or in the victim's hand
- Hair trapped in a cracked windshield in a vehicular homicide (can place individuals in certain seats)
- Hair superimposed on blood or other liquids at the crime scene
- Pubic hairs in a sexual assault where the perpetrator has worn a condom
- Hair recovered from wrappings on the body or the body bag
- Hair recovered at autopsy
- Hair stuck on the murder weapon, tape bindings, or ligatures
- Hair collected from probative locations on clothing (inside underwear)
- Hair in the mouth, throat, vagina or rectum
- Hair in a discarded mask or abandoned getaway vehicle
- Hair adhering to bumper or undercarriage of a hit-and-run vehicle
- Hair from the suspect's vehicle (trunk or passenger seat)
- Hair in the mechanism or tape of an explosive device

In any case, hair samples as small as 2 mm may be used for mitochondrial DNA analysis and eventual comparison to known individuals. Hair as old as four decades has been successfully tested, and it is unknown what the most extreme age for successful testing might be. In general, scientists have observed that full or partial mitochondrial DNA profiles can be obtained most successfully on recently shed hairs, but that hairs over 21 years of age provide partial or full profiles in almost 80% of cases. With this information, cold cases dating back to the 1960s are candidates for re-examination.

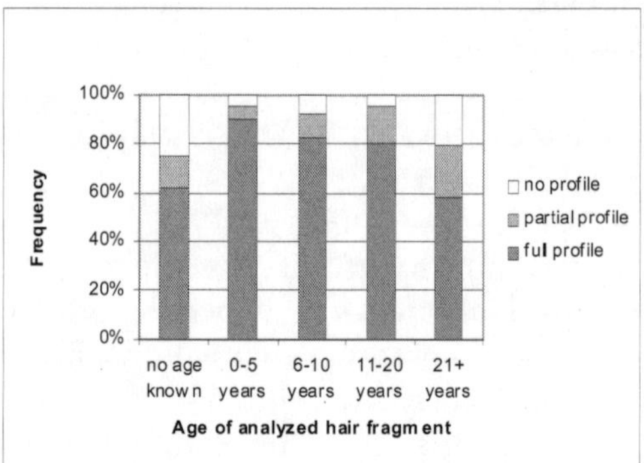

Fig. Older hairs (analyzed many years after a crime) typically are more difficult to analyze than younger hairs, however almost 80% of hairs in their third decade after collection yield partial or full mitochondrial DNA profiles.

In the event that crime scene hairs are located for examination and analysis, several steps leading up to the actual mitochondrial DNA analysis itself will be helpful. While following standard chain of custody protocols, an investigator should plan for a brief microscopic evaluation by a qualified hair microscopist. The hair can be measured and photographed, and if a sizeable plucked root with follicular or sheath material is determined to exist, nuclear (STR) DNA testing can be attempted. Hairs can be slide-mounted for comparison to other hairs in the case. Rather than comparing the questioned hairs to exemplar (known) hairs from reference individuals in the case, which, as stated above, can lead to false positives or false negatives, a microscopist can classify or group questioned hairs according to their general appearances, which then allows the selection of one or two hairs from each group for mtDNA analysis.

On rare occasions, an exclusion of an individual may be obtained without DNA analysis because the physical differences between questioned and known hairs are extreme, but this is an exceptional occurrence. In general, the hairs selected for mtDNA analysis should have substantial probative value to the case, as the expense of testing usually will mean that a limited number of hairs can be analyzed.

Determining the probative value of a questioned hair may be critical to a case. Unlike semen or blood, which is often present due to criminal activity such as sexual or physical assault, hairs may simply be deposited from an individual who has been innocently present before, during, or after the crime has occurred. By themselves, hairs are not indicators of any specific activity, especially since humans naturally lose 75-100 head hairs per day. Exceptions to this are forcibly removed hairs such as those yanked out during a struggle, or pubic hairs, which are not shed onto floors or furniture as often as head hairs. The STR DNA match rate to suspects or victims when semen or blood is present is known to be about 70%; this is due to the fact that semen and blood are often present because of the crime. However, the mtDNA match rate with hairs is significantly lower: about 50%. This is because hairs are everywhere in our environment and may simply collect on clothing, shoes, floors, and furniture and be unrelated to crime scene activity.

Since maternal relatives share the same mitochondrial DNA profile, using mtDNA analysis to investigate a within-family crime is not useful, for example, when one sibling has allegedly been murdered by another. Hairs found at the scene will be uninformative as to the correct donor based on mitochondrial DNA analysis. However, all probative hairs should be collected, for if the theory of the crime later is determined to be incorrect, and a non-family member becomes a suspect, the hairs may become useful evidence.

Hairs found on the floor of a public restroom will be less valuable to the theory of the crime than non-victim hairs found inside the victim's clothing at that location. Because mtDNA is not a unique identifier, linking a hair from clothing to a suspect is more powerful than linking a hair from the floor to that suspect, since an argument can always be made that the hair was contributed by someone unrelated to the case who just happened to have the same profile as the suspect.

The analytical process

A mtDNA analysis begins when DNA is extracted from biological material, such as a tooth, blood sample, or hair. Extraction is the most critical stage of the analysis, because the DNA that is authentic to the sample is being purified away from all other biological materials in the sample, and laboratory personnel are concerned with the integrity and cleanliness of the sample, laboratory environment, equipment, and chemical reagents. Typically, a laboratory carefully cleans a hair shaft in an ultrasonic water bath, and then grinds the hair into a sterile solution. This solution is treated with chemicals to separate the DNA into a new sterile tube.

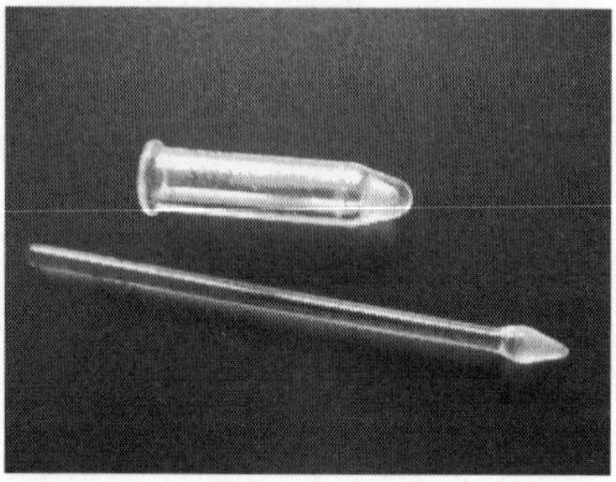

Fig. Photograph of the mini-glass grinder that is used to break up a hair into a chemical solution during the DNA extraction stage. The grinder is effectively a mortar and pestle in which the two pieces are machined to fit perfectly together for grinding.

A process called the polymerase chain reaction (PCR) is then used to amplify, or create many copies of, the two hypervariable portions of the non-coding region of the mtDNA molecule, using flanking primers. This region contains the mitochondrial DNA sequence information: the exact order of the As, Cs, Gs, and Ts that characterize that sample. Primers are small bits of DNA that identify and adhere to the ends of the region one wishes to PCR amplify, therefore targeting a region for amplification and subsequent analysis.

Care is taken to eliminate the introduction of exogenous (contaminant) DNA during both the extraction and amplification steps via methods such as the use of pre-packaged sterile equipment and reagents, aerosol-resistant barrier pipette tips, gloves, masks, and lab coats, separation of pre- and post-amplification areas in the lab using dedicated reagents for each, ultraviolet irradiation of equipment, and autoclaving of tubes and reagent stocks (Figure 6.). Questioned samples are processed at different times than known reference samples and they are usually processed in different laboratory rooms.

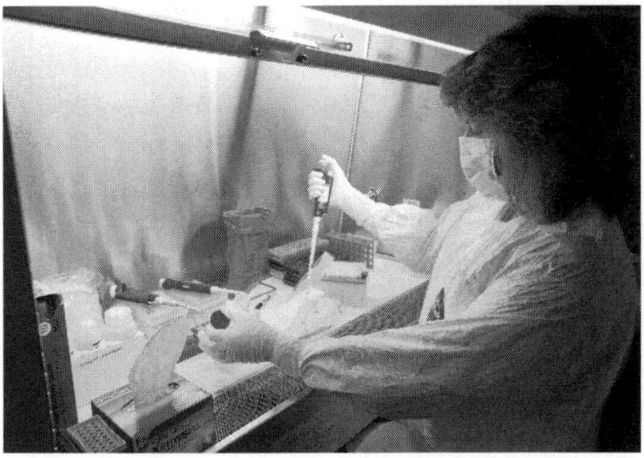

Fig. Disposable equipment, gloves and gowns are required to prevent cross-contamination, carry-over contamination, and sporadic contamination.

When adequate amounts of PCR product are amplified to provide all the necessary information about the two hypervariable regions, sequencing reactions are performed. These chemical reactions use each PCR product as a template to create a new complementary strand of DNA in which some of the As, Ts, Cs, and Gs (DNA bases) that make up the DNA sequence are labeled with dye. The DNA strands created in this stage are then separated according to size by an automated sequencing machine that uses a laser to "read" the sequence, or order, of the DNA bases. Where possible, the sequences of both hypervariable regions are determined on both strands of the double-stranded DNA molecule, with sufficient redundancy to confirm the DNA bases (A, C, G, T) that characterize that particular sample.

At least two forensic analysts independently assemble the sequence and agree on the final DNA sequence that has been obtained (Figure 7). The entire process is then repeated with a known sample, usually a blood or saliva collected from a known individual. The sequences from both samples, about 780 bases long each, are compared to determine if they match. The analysts assess the results of the analysis and determine if any portions of it need to be repeated.

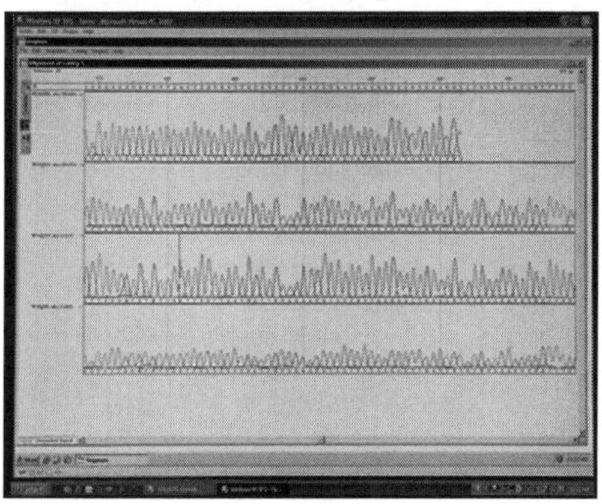

Fig. Example of DNA sequence obtained in a forensic case.

Finally, in the event of an inclusion, or match, the SWGDAM (Scientific Working Group on DNA Analysis Methods) mtDNA database, which is maintained by the FBI, is searched for the mitochondrial sequence that has been observed for the samples.

The analysts can then report the number of observations of this type based on the nucleotide positions that have been read. The number of times that a type or profile has previously been observed is used to calculate a simple statistic which guides understanding of both the court and trier of fact about the significance of the match.

It is important that a mitochondrial DNA analyst state clearly, both in a final report and in testimony, that an individual and his or her maternal relatives cannot be excluded as the donor of a questioned hair. With this statement of a "match", the analyst may qualify the statement with additional information such as "We are 95% confident that the true frequency of this type in North American populations does not exceed 0.06%". In other words, the analyst is saying that there is 95% confidence that at least 99.94% of North Americans will not have the type in question, and does not try to state that the profile is unique.

MITOCHONDRIAL DNA COLD CASES

Lori Roscetti homicide

Lori Roscetti, a medical student in Chicago, was abducted, raped, and murdered in 1986. Four Chicago gang members were arrested for the crime, tried, and convicted in 1988. A review of the case evidence some years later by defense attorneys revealed the presence of numerous undetected semen stains on the victim's raincoat as well as three un-analyzed hairs. STR typing of the semen stains eliminated all the original convicted men and the victim's boyfriend as donors of the semen. Mitochondrial DNA analysis of the three questioned hairs also eliminated all the original convicted men. The convictions of these four gang members were vacated in 2001.

Within one month of the exonerations and release, another individual came forward to report that his brother and an accomplice might have committed the Roscetti murder. Because the informant and his brother were maternally related, the informant donated a sample for mitochondrial DNA comparison to the questioned hairs. This comparison revealed that the informant and his brother were not excluded as the donor of one questioned crime scene hair. The accomplice's mother also donated a sample for mitochondrial DNA analysis.

The results of this comparison showed that she and her son were not excluded as the donor of a second questioned crime scene hair. Based on this preliminary and suggestive mitochondrial DNA testing, court orders were obtained to procure blood samples from the two new suspects. STR typing of their blood samples revealed that the new suspects were likely donors of the semen stains on the victim's raincoat. Both suspects were arrested, and charged, and later confessed to the murder of Lori Roscetti.

Terrorist storage locker

Armenian nationalists rented a storage locker in the Cleveland area during the 1970s for the storage of guns and explosives. Rental fees on the locker were not paid in 1996, which triggered an opening and investigation of the locker by authorities, including The Bureau of Alcohol, Tobacco, and Firearms once munitions were discovered. Several evidentiary hair fragments were collected from a coat and moving blankets inside the locker. Mitochondrial DNA analysis of these fragments in 1999 matched their profile to that of the leader of the terrorist group, Mourad Topalian. Topalian was arrested, charged, convicted and sentenced to 37 months in prison in 2000. An "ancient" mitochondrial DNA analysis was necessary for the hairs, because their mtDNA was minimal and degraded after exposure to the heat of the storage facility over many years. This slightly more specialized approach allows abundant but degraded DNA, such as mtDNA in 25-year-old hairs, to be captured in smaller fragments.

9

Resource Needs of the Forensic Science Laboratory

Introduction

The requirement contained in the Review Terms of Reference to assess the resource needs of the Laboratory can best be approached by distinguishing between the immediate resource needs, given the Laboratory's current functions (including in respect of the limited DNA analysis role accorded to the Lab at the present time) and those which would arise from the introduction of the proposed DNA database. These resource requirements are examined separately in this section and in the next by first quantifying the demand for forensic services in each case and then estimating the staff resource required to meet that demand.

EXISTING DEMAND FOR FSL SERVICES

The number of cases received from the Gardaí and cases reported (i.e. processed) over the past five years. The divergence between cases received and reported, highlights the fact that the Laboratory has traditionally not had the resources to satisfy the forensic analysis demand of the Gardaí. In particular, total unreported cases between 2000 and 2006 amounted to 33% of all cases submitted.

A range of mechanisms are in place to manage this, including provisions with respect to prioritisation contained in the Service Level Agreement with the Gardaí. In addition, it is agreed policy to return cases to the Gardaí where they have not been analysed within nine months of receipt into the Laboratory, which prompts a re-evaluation of the priority to be accorded to the analysis. Furthermore, cases with imminent Court dates are prioritised and every effort is made to ensure that Courts are not delayed by lack of a report from the Laboratory. Similarly, the capacity for the Gardaí to flag a case as having increased in priority also exists.

Notwithstanding these arrangements, the Gardaí would undoubtedly prefer to have the cases they have submitted examined and the only conclusion can be that the detection and prosecution of crime is weakened as a consequence of the unmet demand for forensic services. This is particularly so in drug analysis cases, where a certificate is required for prosecution, and in those cases where DNA analyses could assist an investigation. In fact, DNA is the most recently established section of the Laboratory and cases submitted to it have effectively been maintained at artificially low levels to take account of the limitations on capacity in that section. It is estimated that a high proportion (perhaps two-thirds) of the cases recorded in the Biology Section would increase in value by adding DNA analyses (e.g. in a case where blood spatter analysis is currently carried out, DNA analysis would often also be useful). Moreover there is informal evidence to suggest that Garda investigators believe that DNA analysis could and should be more frequently used in investigating serious crimes.

Submerged demand for forensic services within the Garda organisation is likely to be activated should the Laboratory begin to meet current unmet demands. It is not possible to estimate accurately the extent of this submerged demand, so the conclusion of this Review is that some conservative estimate of that submerged demand should be provided for and, furthermore, that actual demand should be monitored on a frequent basis and that the gap between demand and response not be permitted to open up again. Some suppression of demand is acceptable in the long term, as the rationing effect forces the persons requesting services to prioritise, but this can also be achieved by close case management and the gap which currently exists is so broad as to undermine the confidence of the Gardaí in the service, which could have negative strategic consequences for the application of forensics in investigation.

Cases received		Biology	Chemistry	DNA	Drugs
	2006	756	1009	523	8852
	2005	785	1165	408	8079
	2004	795	1165	387	8192
	2003	765	1268	306	8470
	2002	872	1585	220	8673
	2001	878	1636	300	10715
	2000	771	1390	188	8658

Cases Reported		Biology	Chemistry	DNA	Drugs
	2006	471	634	378	7248
	2005	419	718	317	6390
	2004	393	681	271	6251
	2003	430	686	249	5653
	2002	528	1188	201	4764
	2001	607	1360	225	5813
	2000	472	1027	138	5111

Table. Total Cases Received and Reported 2000-2006

	Biology	Chemistry	DNA	Drugs	Total
Received	5,622	9,218	2,332	61,639	78,811
Reported	3,320	6,294	1,779	41,230	52,623
Short fall	2,302	2,924	553	20,409	26,188

ADDITIONAL SERVICES NOT CURRENTLY BEING PROVIDED

In addition to the unmet demands for the existing services available from the Laboratory, the range of services supplied is not comprehensive when compared with other forensic Laboratories and the Laboratory is under growing pressure to provide these additional services.

Two notable examples include the provision of more extensive services in toxicology and in digital evidence. In the past the Laboratory has carried out alcohol determinations on biological samples submitted relating to victims of sexual assault cases. In more recent years these requests have been expanded to test for the presence of controlled drugs and the Laboratory has not been able to meet this need. If a service were to be provided, it is estimated that approximately 150 of such cases would arise annually.

The experience of other Laboratories particularly in Northern Ireland and rest of UK is that there has been a dramatic increase over recent years in requests for examinations of information stored on mobile phones. In the Northern Ireland forensic lab for example, 5 staff are employed full-time in providing this service. Thirty seven such requests were received in FSL in 2006 but if a more comprehensive service were offered it is believed that significant submerged demand would be activated.

BENCHMARKING RESOURCE NEEDS IN FORENSIC SCIENCE

In order to evaluate the staffing resources which would reasonably be required to meet the existing demands, an exercise was carried out which compared the productivity of the Irish Lab with a number of leading comparable European organisations.

Fortunately, as the timeframe for this report's completion would not have permitted entirely original research on this complex subject, a well-established benchmarking model, QUADROPOL, was already available. This project was sponsored by the EU OISIN II programme and published in 2003. It compares the productivity of four European Forensic Science laboratories, the Finish National Bureau of Investigation Crime Laboratory (RTL), the Dutch Nederlands Forensisch Instituut (NFI), The Polish Institute of Forensic Research (IFR) and the Swedish National Laboratory of Forensic Science (SKL).

For FSL and three of the QUADROPOL laboratories. The data for the Polish laboratory have not been used since that laboratory's tasks differ significantly from that of a national forensic science laboratory. It should be noted that FSL does not compile data on the type of cases in exactly the same manner as the QUDROPOL project and thus it was not entirely straightforward to derive the FSL data. There are also important differences between each of the laboratories studied concerning the number of cases in a discipline area, the scope of those areas, as well as the level of automation and computer support. Thus the figures given are not possible to compare directly in all cases. However, they can certainly give an estimate of the comparative productivity of FSL from an international viewpoint.

Table. Productivity Data

	FSL 2006		Cases per FTE:S				Hours per item			
	Prod cases	FTE:s	FSL	SKL	NFI	RTL	FSL	SKL	NFI	RTL
Firearms	39	1.5	26	132	74	61	9.5	1.0		4.4
DNA	376	16.3	23	128	59	231	2.1	2.8	4.2	1.5
Paint and glass	168	4.6	37	62	79	79	3.0	2.0	6.2	1.9
Drugs	7334	17	431	429	381	230	1.3	0.8	0.8	1.3
Marks	44	1.3	34	71	44	88	6.4	2.2	3.8	2.9
Hairs and fibres	50	2.5	20	155	44	54	2.3	1.2	4.6	3.6
Fires	209	1.5	139	152	79	126	3.2	2.5	8.8	1.8
For tox	82	0.9	91		373					
Digital ev	37	0.25	148	9	15		15.2	30.1	15.2	
Explosives	14	0.5	28	90	27	37	5.1	2.7	6.0	6.0

- *Paint and Glass, Marks, Hairs and fibres, Fires, Explosives*: The production conditions may be assumed to be similar.
- *Firearms*: FSL performs only SEM analyses while the QUADROPOL laboratories also include ballistic measurements.
- *DNA*: Sweden and Finland (SKL and RTL) both have DNA registers. As a consequence a large proportion of the cases handled are more straightforward volume crimes which may account for the high numbers of cases per FTE. FSL uses only DNA analyses for more complicated cases where more items are sent in per case.
- *Drugs*: FSL only performs identification and quantification cases while several of the QUADROPOL laboratories also do more time demanding profiling cases.
- *Forensic toxicology*: FSL at present performs screening analyses in a relatively small number of cases while NFI has a complete service using more automation.
- Digital evidence FSL investigates only mobile phones while SKL and NFI has a broader investigation area

In comparing the data account must be taken of several factors which affect recorded productivity. Productivity is impacted by the mix of cases between those requiring extensive work and those that need comparatively little work. Since FSL prioritises the most important cases (and does not reach all cases), it is likely that it handles a greater proportion of cases that are more timeconsuming. Another issue is the demands of the Courts on the organisation, which has been remarked upon. It was quite clear in extracting data for the above exericse and from the review of the FSL operations generally that the Court and procesution system is much more demanding on resources than those in the Swedish and Finnish laboratories, for example. In addition, and again as referred, the FSL experts devote considerably more time to the recording of evidence due to the requirements of the prosecutorial process and the relatively limited reliance on digital recording.

Taking all of the above into account, one can infer that the productivity is good in some areas while it can be improved in others. The overall position is that the Laboratory can be said to compare well with these comparable organisations. While there is scope for improvement, this is only likely to be achieved in the medium to long term, through improved computer support incorporating closer cooperation with the Gardai as well as greater automation in the drugs and DNA areas. Reform of the restrictions imposed by the Irish criminal justice process would also yield results, but again this would take time to achieve and requires further study.

Resource Needs of Current Functions

Broadly speaking, then, the comparison with the QUADROPOL study illustrates that FSL is operating at productivity levels akin to those in leading European labs. with some additional burdens deriving from factors particular to the court process in Ireland and limited automation of processing or of information management between the Lab and the Gardaí.

As the first of these factors is largely outside administrative control and the second will take some years to enhance, it is reasonable to assess resource needs by reference to the existing productivity levels. On this basis the resource requirements can be calculated as follows as set out in Table below;

Area	Cases Rec'd 2006	Cases Reported 2006	Suppressed demand	FTEs Present	FTEs Sanctioned	Needed Current Demand	Needed Total
Biology	756	471		12.8	16	20	20
Chemistry	1000	634		14.7	18	23	23
DNA	523	378	1000*	7.5	14	10	20
Drugs	8852	7325	10000*	17	21	20	23
New Services							2
Total	**11140**	**8808**	**11,000**	**52**	**69**	**73**	**88**

The Review's finding then is that in order to meet the current unmet demand at total of 73 scientific staff are required (4 more than currently sanctioned). To meet estimates of suppressed demand for DNA alone, this total should be increased to 83. If conservative estimates of suppressed demand for drugs are taken into account, a total of 86 required. The Review also estimates that 2 further staff would also be required to meet the demands for new services.

ADMINISTRATIVE STAFF REQUIREMENT

It was evident during the review that the administrative support capacity of the Laboratory is also under significant pressure, particularly in respect of its case intake function. This is currently staffed by one HEO and 2 EOs. The Review found the work of this area to be an extremely busy one, and very demanding administratively. Its chief functions are to receive and log case material delivered by members of the Gardaí, track cases to sections within the Lab and subsequently return items to the Gardaí.

To these core functions, however, must be added a significant workload arising from the need to give evidence at Court to certify continuity of case handling. When analysed it was found that at any given time, at least one of the 2 EOs was attending court and with normal absences for leave etc., the consequence is that scientists are regularly called on to cover the case intake and return function, which is highly wasteful of this specialist resource. The appropriate level of cover for the case reception area, based on the volume of work received and the time required to process this, is believed to be two EOs at all time. To ensure that this can be provided on a continuous basis, without routinely requiring cover from scientific staff, it is recommended that an additional two EOs be assigned. It is also recommended that an additional CO be assigned to allow for a dedicated data entry and other support services.

OVERVIEW OF INTERNATIONAL DEVELOPMENTS IN FORENSIC SCIENCE MANAGEMENT

Forensic science has become an increasingly important element in criminal justice systems throughout the world. The requirement for forensic evidence has grown internationally, causing the caseload of many laboratories to double or treble in recent decades. There has also been rapid development of new techniques and methods giving improved possibilities to obtain forensic evidence. At the same time the demands on the laboratories from customers (mainly the police) and the financing bodies have risen. There is pressure to improve timeliness and provide more "science for money".

There is also increased recognition that forensic science should be an integral part of the investigation and justice process. The work of a forensic laboratory does not only concern evidence for the courts but should also give impulses to the investigation process. The findings of the laboratory can often give valuable information that can assist the investigation or can be used as intelligence. Thus there is a need to improve the co-operation between the forensic science laboratory and the police and to improve the forensic awareness in all levels of the police.

With increasing pressures on the forensic scientists to produce more within shorter timeframes the risk of failure has increased. International best practice has been to manage this risk by introducing quality assurance systems and accreditation from appropriate external bodies. Many laboratories have also understood that with the increased demands and laboratory size there is also a need for competence assurance systems and important developments on this front are being championed by the European Network of Forensic Science Institutes (ENFSI).

Another noteworthy area of development is the increased use of IT and automation in order to rationalise workflow and improve its quality. In several countries systems have been developed where the information from the crime scene and the police is transferred in digital form to the laboratory where all the analyses results and other laboratory information is added. The information that is pertinent for the laboratory report is transferred by the system to form a preliminary report that is amended by the expert. The report is sent in digital form with a digital signature to the police and can then be forwarded to the prosecutor and the court.

Automation has been introduced in areas where a large number of samples are analysed by the same method. Automation greatly limits the impact of human error in the analysis and has been especially prevalent in the areas of drug and DNA analysis.

During recent decades there have also been important developments in how forensic evidence is evaluated. A new approach has emerged, beginning in the UK but also spreading to many other European countries, especially those having adversarial justice systems. The details of this new perspective are outside the scope of this review but in simple terms the new approach weighs the likelihood of a particular forensic result, based on stated hypotheses about the case, rather than in a supposedly 'pure' scientific vacuum. The significance of this approach, which is increasingly seen as giving more meaningful evidence, is that precisely because the analysis in not taking place in a vacuum, greater dialogue with all stakeholders in the justice system is necessary.

An important driver in the development of forensic science internationally has been international co-operation and all forensic science laboratories need to dedicate resources to keeping abreast with international developments. Resources are required to participate in international forensic science co-operation, such as that under the aegis of ENFSI and in associated R&D. Similarly, resources are required to invest in training, not merely within the forensic laboratory itself, but also with other stakeholders generally.

CURRENT STATUS OF THE IRISH FORENSIC SCIENCE LABORATORY (FSL)

The Laboratory was established in 1975 to provide impartial scientific evidence following examination of items from crime scenes. Its mission, as set out in its Strategy Statement is to 'assist in the investigation or crime and to serve the administration of Justice in an effective manner by a highly trained and dedicated staff providing scientific analysis and objective expert evidence to international standards.' This core objective is complemented through the implementation of a range of initiatives in keeping with the public sector modernisation programme generally.

Organisation and Staffing

The Laboratory has grown organically with additional staff numbers being allocated on an ad hoc basis. Work is carried out in four functional areas – biology, chemistry, drugs and DNA. The staffing situation is summarised at Table below.

Grade	Number of Authorised Posts	Vacancies	Number of FTEs
Director	1	0	1
Deputy Director	1	0	1
Grade 1	10	0	10
Grade 11	21	0.4	20.6
Grade 111	20	4	15 (16-1 on career break)
Senior analyst	8	0.4	7.6
Analyst	12.5	11.5	1
HEO	1	0.3	0.7
EO	3	0.2	2.8
CO	3	0	3
Total	**80.5**	**16.8**	**63.7**

It became clear during the course of the review that, despite the fact that staff had been sanctioned and that recruitment panels actually exist in some instances, it nevertheless appeared to take an indefensibly long time to finalise the appointment of new staff. These delays are serious and need to be seen in the context of the extent to which the Laboratory is currently unable to process all Garda submissions, as examined in Section 3 of this Report. Recruitment of staff to fill the existing vacancies therefore clearly needs to be accelerated, and if the recommended staff increases proposed in Section 3 are approved, special measures should be put in place to bring the extra staff on stream as quickly as possible.

STAFF AND CAREER PATHS

A majority of staff within the Laboratory are scientifically qualified, many to advanced postgraduate level. Retention of staff is high and the typical profile is someone who joins the Laboratory mid to late twenties and stays for the rest of their career. The limited career progression is naturally a subject of concern and while detailed proposals are outside the scope of this review, it clearly something to be factored into Human Resource management decisions.

Case Processing in the Laboratory

Routine cases are received into the Laboratory and stored before allocation to a scientist in a particular section. The allocation is carried out by the Section head based on the importance and urgency of each case. A fixed number of high profile cases are taken in directly by the scientist who keeps responsibility for that case and acts as liaison between the Gardaí and the Laboratory. Typically, a large proportion of items are submitted with minimal analytical request information. At the submission stage it is assumed that the scientist will carry out the most appropriate analyses suited to address the question raised by the allegation. There is a lot of communication between the investigating officer and the scientist. At the beginning of a major case it is not unusual to have multiple phone calls from different members of the Gardaí. Recent understandings with the Garda authorities aim to have all the communication between the scientist and the senior investigator.

The situation relating to Drugs is different. Samples are submitted in tamper evident bags and analyses needed are obvious. However, even in relation to Drug cases, there is a lot of manual handling and duplication of data entry. The receipt of items into the Laboratory is very cumbersome. This is due in part to legal constraints and in part to practices grown with the development of the Laboratory.

Samples are physically received with a Garda C56 form which lists the items being submitted. The practice has developed that labels on the packing and detailed descriptions of the contents form part of the Laboratory report. This is a time consuming aspect of the work and should be rationalised but remains necessary because it is not unusual for the questioning in court to focus more on the physical appearance of the items rather than on analytical results. This is true even in Drug cases where the court may want to probe what the Garda recorded or saw prior to submitting the sample to the Laboratory.

The review found that the inadequacies of case handling need to be addressed at at least three levels. In the short term, FSL and the Garda authorities need to review the arrangements for case management to ensure they are as streamlined and effective as possible. Over the medium term, however, the transmission of information needs to be automated further, IT solutions which properly communicate with each other are required and manual entry of data needs to be kept to the minimum.

Beyond the operational practices of the FSL and the Gardaí on this front, however, it also became very clear during the review that the extent of the requirements arising from the need to certify the chain of evidence in an Irish context was generating enormous and disproportionate demands on the Laboratory and others involved in the criminal justice process. Indeed, just as questioning in Court often focuses on the appearance of evidence, there is also a significant emphasis on the chain of evidence as opposed to the validity of the findings themselves.

It was clear from the examination of the Swedish work practices that systems could be put in place which certified the chain of evidence in a common-sense fashion but which did not require routine attendance at Court and yet there was no suggestion that the judicial process was less fair or robust as a consequence. Previous attempts to minimise this demand in an Irish context appear to have failed and it is therefore recommended that the Law Reform Commission or some other suitable expert body be asked to study the practical application of the law in this regard and evaluate the scope for a more efficient means of approaching the management of physical evidence.

QUALITY ASSURANCE AND ACCREDITATION

The Review found that the Laboratory has aimed to be as up to date as possible in respect of quality assurance, in spite of limited resources. In particular, a successful application for accreditation to ISO 17025 standard was made in 2002. The approach adopted was to gain accreditation for one particular analytical method a year in each of the main sections. This process has now advanced to a point where most tests in routine use are accredited. Due to the greater variety of test methods in the Chemistry section a number of areas remains to be accredited and the Review recommends that this process be completed within three years, which represents a reasonable timeframe for this purpose.

Insofar as competency testing is concerned, training records are maintained as part of the overall quality system and staff are required to carry out actual tests to prove competency. In keeping with the international trend in this regard, however, and with the best practice being promoted by ENFSI it is likely that more systematic competence testing will be examined by the Laboratory. The Review found that such an approach would be particularly beneficial, if not actually essential, for the Laboratory to expand as recommended in this review.

SOURCE DESIRES OF THE FORENSIC SCIENCE LABORATORY – INTRODUCTION OF THE DNA DATABASE

A similar approach was adopted in order to assess the resources requirements of the DNA database, beginning with an assessment of the likely volume of samples which will require processing.

EVALUATION OF THE VOLUME OF SAMPLES LIKELY TO BE SUBMITTED TO THE DNA DATABASE

In summary, two key streams will require analysis - 'suspect' samples and crime scene samples. Suspect samples will be taken from persons detained for arrestable offences or from person in prison on the commencement of the legislation. Crime scene samples will be determined by operational practice and it is expected that this can be further divided into the very serious crime scene cases which the Lab already provides a service for and higher volume crime, typically burglaries and unauthorised taking of motor vehicles. In the model developed in Appendix A it is assumed that the latter category begins at very modest levels and expands gradually as crime scene procedures are developed in response to the availability of the database. There are a number of other potential streams of sample generation under the legislation including volunteered samples in mass-screening situations, elimination samples from law enforcement staff, crime specific samples and finally the 'second' samples taken for evidence purposes when a positive intelligence hit is obtained from the system. These are all taken into account in the proposed model.

Resources Required – Pre-commencement

Considerable preparation will be required in advance of the DNA database being introduced. During this preparatory stage procedures will need to be developed following study of other laboratories that have a working register. Personnel will require to be trained, instruments purchased and computer support developed.

It is estimated that a team of up to 6 staff will be required for this phase. Included in this team will be a person to lead and take overall responsibility for the project. Given its significance it is envisaged that this person should be at Deputy Director level and assume overall responsibility for DNA services in the Lab. They will need to be assisted by at least one senior scientist to oversee the detail of procedure development and training as well as an IT Manager to develop the process automation and the systems to support the eventual database itself. Other key responsibilities for this team would be to develop a procurement programme for equipment and consumables and to develop liaison mechanisms with the Gardai specific to the needs of the database's operation.

Resource Requirement Post-Commencement

The process of modelling this requirement is quite complex and driven by:

- The number of items requiring testing according to case type
- The number of tests required per sample
- The time required to carry out a test
- The time required to carry out subsequent analysis.

The number of items generated per case will vary and is estimated in this model based on current experience with other types of submissions to the Laboratory. The more serious crimes will be assumed to generate the most samples. Depending on the sample type, either one or two tests will be carried out of each item submitted. For suspect samples, two tests will be carried out. For crime scene samples one test will be carried out as the comparison against the database and any suspects will form a de facto cross-check in any event.

Based on an examination of the processes applied by the Swedish Lab, an estimate has then been carried out of the time required in hours to test each samples. The model assumes greater time commitment in crime scene samples where the stain has to be examined and a sample extracted than in more straightforward suspect sample, which will usually be readily analysable mouth swabs. In all cases the model assumes productivity growth over time as processes are embedded and enhanced. Finally, the analysis time for each type of sample is also examined, again by reference to the SKL experience and in the light of FSL's existing experience in this field.

The net output of this model identified a requirement in terms of staff hours and, assuming operationally available processing time of 70% of FTE staff time, this would give rise to a staffing requirement of 64 in the first year. It should be noted that this would include the existing staff complement of 35. For the purposes of this analysis, certain functions currently classified as being in the biology field are included in the complement for the DNA database, given their significant contribution in the extraction of DNA from crime sample extraction.

The model also suggests an increase in the staffing requirement over the first five years of operation, as DNA is submitted from a greater number of crime scenes. More detailed implementation analysis will be required to assess how quickly the system moves from its start up volume, given the scale of what is contemplated and the complexity of the processes and interactions with the Gardaí which will need to be developed.

NON-STAFF RESOURCE REQUIREMENTS

General Equipment and Accommodation Needs

FSL is reasonably well equipped in terms of scientific and analytical equipment in order to deliver on its current mandate. In view of the manner in which the Laboratory has evolved over time, however, equipment is acquired and replaced in a relatively ad hoc fashion and the approach to this issue would benefit from a planned replacement programme, particularly insofar as more expensive machinery is concerned. The current FSL accommodation is completely inadequate, as is the provision for sample storage. However, as plans are at an advanced stage to build a new facility for the Lab, this issue was not studied in any detail as part of the Review.

DNA Database

The introduction of the database will give rise to non-pay expenditure under the headings of DNA Analytical Equipment, DNA Analysis Consumables and IT system management. A detailed costing exercises is outside the scope of this review but an indicative outline of the likely costs is set out at Appendix D. Based on the Swedish experience in particular, the total capital cost is estimate at being in the region of €3m, including very tentative €2m for the IT component. Substantial consumable costs will also be incurred on an ongoing basis (primarily sample test kits).

LIAISON WITH AN GARDA SIOCHANA

Current Situation

The relationship between An Garda Síochána and the Forensic Science Laboratory spans several decades and has evolved into a partnership approach. The existing relationship is governed by a Service Level Agreement (SLA) signed between the Commissioner and the Director of the Forensic Science Laboratory. There is also a quarterly meeting of the Advisory Forum jointly chaired by Assistant Commissioner, National Support Services and the head of Chemistry Forensic Science Laboratory. An Garda Síochána has placed the effective use of scientific and technological advances at the core of its Corporate Strategy in contributing to improved detection rates and crime reduction. This is designed to complement the current reorganisation of investigative processes to provide demonstrable levels of responsibility and accountability in the end to end process of crime investigation.

It is now recognised that the effective use of forensic evidence is dependent on forensic investigation activities being integrated into the investigative process from the outset of an investigation. To maximise the outcomes mechanisms and structures need to be in place where this integration takes place as a standardised norm rather than happening in exceptional or ad hoc circumstances. The establishment of the DNA database will incorporate legislative provision for the establishment of an oversight body to monitor the operation of the database and report to the Minister. The committee will have responsibility to review the arrangements and codes of practice for the taking of samples (including crime scene samples), their transmission to the laboratory, their secure handling and storage, the development of profiles and the communication of the results of analysis.

The current review affords an opportunity to develop a framework to ensure optimum usage and delivery of forensic services. It will provide the basis on which to develop codes of practice and protocols required by legislation. The core of the proposal is to establish an effective two way communication flow and provide a mechanism to facilitate mutual accountability and demonstrable transparency in the processing of DNA samples.

Proposed Framework for Enhanced Liaison Arrangements

It is envisaged that each of these levels would inform the activities and progress of the other to ensure adequate and effective levels of service are delivered and maintained.

Corporate Agenda

It is proposed that Forensic Investigation Strategy is prioritised on the corporate agenda of An Garda Síochána. Specific prioritise will be set out in the terms of the Service Level Agreement. This will assist in determining the strategic direction of forensic integration and prioritisation of requirements to achieve the desired outcomes on behalf of An Garda Síochána (the customer). These priorities can then be reflected in the strategy statement of the laboratory to ensure delivery.

Regional Forensic Investigation Seminars

Deriving from the above each Regional Commissioner will chair an annual Regional Forensic Investigation Seminar. This will provide a forum at which to discuss forensic issues at two levels strategic/tactical and operational. The meeting will be attended by the Senior Management Team of the Region, senior laboratory scientists and Garda Technical Bureau Management.

The Detective Superintendent in consultation with the regional management team will set the agenda and will focus on specific items relating to forensic investigation relative to the region. At a managerial level the focus will be on the operation of the Service Level Agreement in meeting the operational policing requirements of the Region. Sergeants in charge of each of the Divisional Crime Scene Units will also attend to address specific issues arising.

The operational focus of the seminar will provide a forum for Divisional Scenes of Crime Examiners, Forensic Science Laboratory scientists and Technical Bureau experts to sharing of best practice processes and techniques. It will also address training issues specific to the units in attendance and allow benchmarking of national practice. The outcomes of the above meetings will set the main theme of the agendas for the quarterly advisory forums.

Advisory Forum

The existing Advisory Forum will continue to meet on a quarterly basis to oversee the workings of the Service Level Agreement. However it is proposed that the structure of the forum be revised to reflect the issues arising at a tactical and operational level in the context of delivering on the SLA. In this regard the outcomes of the Strategic/Tactical and Operational meetings will inform the agenda.

In addition to monitoring the Service Level Agreement, the advisory forum will be in a position to advise the Commissioner on the ongoing implementation of a Forensic Strategy within An Garda Síochána to ensure optimum use of advances in science and technology.

FORENSIC MANAGEMENT TEAMS (FMTS) IN MAJOR INVESTIGATIONS

The introduction of Major Crime Investigation Teams and Senior Investigating Officers (SIO)s necessitates a structured and standardised approach to the integration of forensic investigation into the investigative process. This can be best achieved through the establishment of forensic management teams (FMTs), supported by a forensic strategy for all major investigations agreed as policy by the Commissioner. The FMT will comprise of the SIO; the Garda Technical Bureau expert and the Forensic Scientist assigned to the case. The team will provide the communication channel to and from the investigation. The team will also advise on the most appropriate use and prioritisation of exhibits to the investigation team.

Crime Scene Co-ordination Unit

There is a statutory requirement enunciated in the Draft DNA Legislation 2006 for An Garda Síochána and the Forensic Science Laboratory to develop Codes of Best Practice. This unit will be tasked with Management of An Garda Siobhan's functions in the establishment of a DNA database including the development of Codes of Practice for taking DNA samples, transmission, storage and outcomes.

Management information and performance indicators need to be developed to allow monitoring of the value and effectiveness of forensic inputs. The unit will be tasked with developing structured mechanisms to co-ordinate and manage forensic outcomes to aid investigators. This will also provide timely management information to assist national and local policy interventions.

Central to the effectiveness of the above is the provision of operational support, appropriate training and monitoring of the activities of Divisional scenes of crime examiners. This support can be provided jointly by An Garda Síochána and the Forensic Science Laboratory through the provision of regional training seminars, accredited training and sharing of best practice.

TRAINING

A forensic investigation and awareness training strategy needs to be delivered at various levels throughout the Garda organisation and the Forensic Science Laboratory.

- A joint training committee under the auspices of the advisory forum should be established to agree a joint training strategy for inputs to various in-house and external training courses.
- Accredited training for scenes of crime examiners with joint input to training model from the Garda technical Bureau and the Forensic Science Laboratory,
- Forensic awareness training at all levels from student training to senior management development courses.
- There should be joint training forensic evidence awareness inputs from the Forensic Science Laboratory and Garda Technical Bureau to all SIO, Incident room managers, Major Crime Investigation Teams etc at appropriate level.

- Reciprocal awareness seminars for Laboratory staff to be provided by Garda Investigators.
- Garda Technical Bureau experts and Forensic Science Scientist should meet twice a year to share and exchange best practice and develop the concept of a 'community of practice' between both organisations.
- Members of both organisations should engage in joint training opportunities and attendance at appropriate seminars and conferences.

General

The effective use of DNA as a tool in the fight against crime is critically dependent on processes within An Garda Síochána. The integrity and reliability of crime scene samples, donor and suspect samples throughout the chain of retrieval and submission must be upheld to ensure integrity. This will impact significantly on Garda processes from the arrest and detention of prisoners, crime scene examination and subsequent submission of samples to the Garda Technical Bureau and their onward transmission to the Forensic Science Laboratory. In the context of reliable evidence there is a co-dependency on processes conducted throughout the exhibit cycle within An Garda Síochána and the processing of samples in the laboratory.

The legislative provision for the drawing up of Codes of Practice will address the logistical arrangements that will be put in place. However this must be underpinned by a commitment to providing the relevant resources to An Garda Síochána. In particular the integration and prioritisation of requirements to align with the current ICT strategy and provide technical and IT based solutions to managing critical processes. The issue of accommodation must also be addressed to provide fit-for-purpose facilities to ensure handling and processing of exhibits is maintained to the highest standards.

The material for the screening RIA submitted to the Department of Justice Equality and Law Reform highlighted these issues. To ensure smooth integration and implementation of the database, it is recommended that these issues be sequenced and addressed by and inter-agency implementation

PROMOTING FORENSIC SCIENCE IN IRELAND INCLUDING THE SCOPE FOR ENHANCING ACADEMIC LINKS AND FOSTERING JOINT RESEARCH

The work of the Forensic Science Laboratory depends on some of the most complex scientific theories and on the use of sophisticated scientific apparatus. In a rapidly changing scientific environment the FSL must meet the highest standards of accuracy in the preparation and presentation of evidence, it must keep abreast of fast moving sciences (eg genetics, chemistry and information science), and it must maintain and enhance its reputation for professionalism at home and abroad. This reputation is especially important within the international profession of forensic science, with other professions with which the staff work closely, eg the legal and medical professions, and with members of the Gardai. Needless to say the public must have confidence in the FSL. The courts, legal community, the gardai and public generally need reassurance that forensic science in Ireland is world class standard.

From 1975 to 2006 forensic science has been used in a significant number of major criminal cases in Ireland. However, comparisons with other countries - UK, Sweden, Finland and Netherlands - show that the FSL has not been able to use science in criminal investigations to its full potential. This has to change. It is extremely important to raise the scientific profile of the FSL in the mind of the complex range of stakeholders with whom it engages.

Benefits from Strategic Links with Academic Science

In its reliance on the use of science, the FSL should be compared with medical, industrial and agricultural laboratories. Like all state-funded laboratories it has been under great pressure to keep up with the advances in science. The Review found that other organisations in a similar position had benefitted by developing links with scientific faculites. Leading hospitals and their scientific and medical staff, for example, have developed closer links with the medical and science faculties in the five university medical schools, supported by the Department of Health and Children and by the Department of Education and Science.

This approach is in keeing with the Government's strategic commitment to build a "knowledge-based society" and reflected in the estalishment of Science Foundation Ireland, and in other initiaitves, such as increased funding from the Department of Health and Children to Health Research Board to support developments in health care and funding by the Department of Education and Science to establish the Irish Research Council for Science, Engineering and Technology (IRCSET). The Department of Agriculture has likewise funded new research programmes.

It is thereofre recommended that the FSL build such academic links, with funding and support from the Department of Justice, Equality and Law Reform. There are multiple benefits to this approach: it will make it easier to remain up to date with the latest developments in the relevant sciences, to build up personal links between the staff of the FSL and leading Irish scientists, to carry out or participate in research at the forefront of forensic science, to encourage staff of the FSL to initiate research projects, to better integrate the staff of the FSL in the Irish and international scientific community, to provide sabbattical opportunities for staff of the FSL to get experience of new techniques and new ideas in leading Irish laboratories, to attract excellent young scientists into forensics through "internships", and in overall terms to enhance the reputation and raise the public profile of forensic science and the FSL in Ireland and abroad.

Establishment of a Lectureship in Forensic Genetics

In order to anchor the development of these links, it is recommended that the Department should establish a funded lectureship, by means of public competition, and with the assistance of the Laboratory. In order to maximise the benefit to the Laboratory from such an exercise, this lectureship should specifically focus on the field of genetics, in light of the increasing centrality of forensic DNA analysis to the work of the Laboratory and having regard to the forthcoming establishment of the DNA database.

The precedent for Government Departments funding lectureships is well established in the Health and Agriculture field and the benefits of professionals having access to sources of academic learning and collaborative research is well established in Ireland and internationally. The desired result is the promotion of a research school which in addition to carrying out fundamental research will act as a scientific resource to the Forensic Science Laboratory whose main function will remain operational casework on behalf of the criminal justice system.

The specific terms of reference for such a competition would be for a matter for further detailed consideration but in general the intention would be that the successful applicant would;

- Be a member of staff of a major academic research and teaching department devoted to the relevant science;
- Carry out high quality research at the forefront of forensic genetics;
- Create an international reputation for Irish forensic science;
- Provide a link between the staff of the FSL and the Irish scientific community;
- Build links between the FSL and the main academic and scientific institutions in Ireland (the universities, Royal Irish Academy, Science Foundation Ireland, etc etc)
- Collaborate with the staff of the forensic science laboratory in research and teaching;
- Organise scientific courses, seminars and conferences for forensic scientists;
- Cooperate with staff of the FSL who want to supervise research outside the FSL, or who want to take sabbatical leave to carry out research;
- Organise information and education programmes for lawyers, judges and members of AGS;
- Participate in public discussions on forensic science;
- Teach the subject of forensic genetics to undergraduates and postgraduates;
- Supervise internships for students interested in becoming forensic scientists;
- And generally enhance the reputation and raise the public profile of forensic science and the FSL in Ireland and abroad.

10

Forensic Science Administration in Context

The environment for forensic science administration is changing. Managers and directors of crime labs must adopt new and higher standards of professionalism, verifiability, and scientific rigor. The knowledge and skills of lab managers must now go well beyond familiarity with scientific techniques of forensic testing. Lab managers must be up to date on changes in the legal, political, and social environment of forensic science. We also need research on how lab managers and the criminal justice system might respond to such changes. Recent events creating these changes include the advent of DNA typing, changes in the rules governing expert witnesses in our federal courts and many state courts, renewed opposition to the death penalty, and recent revelations of failures of forensic science administration in the Houston Police Crime Lab, the Seattle Crime Lab, the FBI explosives unit, and other centers of forensic science.

DNA TYPING

DNA typing gained admission to US criminal courts only when was shown that the technique had passed high standards of scientific rigor. This significant advance in forensic science caused judges and lawyers to acquire a heightened awareness of the nature and importance of scientific standards in the testimony of expert witnesses (Faigman, 2001).

At first, the courts were reluctant to admit DNA evidence. This then-strange technique could not gain admissibility until its proponents could show that it was fully scientific and highly reliable. The standards created in that process have been applied with increasing regularity to other forensic techniques including bite-mark analysis, which is more or less discredited, and fingerprint identification which is coming under increasing scrutiny. Forensic techniques that cannot be shown to satisfy the new standards of scientific rigor set by DNA, including fingerprint identification, are at risk of being excluded from American courts.

RULES OF EVIDENCE

Shortly after the advent of DNA typing, the Supreme Court of the United States issued a new set of guidelines for the admissibility of expert witnesses in its "Daubert" decision of 1993. In this case, "William Daubert et al. v. Merrell Dow Pharmaceuticals, Inc.," the court adopted a set of guidelines for admissibility of expert testimony. (The case involved a drug meant to relieve pregnant women of morning sickness, not anything relating to DNA typing.) While Daubert v. Merrell Dow was a civil case in Federal court, the "Daubert" guidelines have been applied in criminal court at the Federal level and in many State courts. The Supreme Court cautioned against reading a checklist out of its decision, but lower courts have often done just that.

The Daubert checklist asks:

- Whether the theories and techniques employed by the scientific expert have been tested;
- Whether they have been subjected to peer review and publication;
- Whether the techniques employed by the expert have a known error rate;
- Whether they are subject to standards governing their application; and
- Whether the theories and techniques employed by the expert enjoy widespread acceptance.

This change in the legal environment of forensic science is extremely important, in part because it has brought ideas from philosophy into the courtrooms of America for use in judging evidence. We believe the forensic science community has not responded to this change in a fully coherent and organized way. As Goodstein (2000) seems to suggest, the philosophy of science appealed to in Daubert does not reflect scientific practice very well. It has become old fashioned and outdated. Nevertheless, it has been read into the Federal rules of evidence and the rules of evidence in many state courts.

Training in the philosophy of science has not been a part of the standard training of forensic scientists or managers and directors of forensic labs. But when forensic scientists enter our courts, they are increasingly likely to be asked to defend the "scientific" nature of their testimony and the criteria of scientific-ness are likely to be the broadly Popperian criteria evoked in Daubert.

THE POLITICS OF THE DEATH PENALTY

DNA typing has led to the exoneration of many prisoners on death row in the US. These exonerations have added momentum to the political movement against the death penalty. Some participants in this movement have argued that current standards of forensic evidence are too low and that "junk science" from crime labs is causing a needlessly high rate of false convictions in our criminal courts. The work of the Innocence Project, headed by the famous defense attorneys Barry Scheck and Peter Neufeld, has been very important for this movement. The Innocence Project is a non-profit legal clinic and not a political group. Their work in exonerating innocent persons wrongly convicted, however, is often cited by opponents of the death penalty.

It may be that the opponents of the death penalty are mistaken. Dezhbakhsh, Rubin, and Shepherd (2003) provide evidence that capital punishment has a strong deterrent effect. It may be that such a deterrent effect is nevertheless not a sufficient argument for capital punishment. Such matters form no part of forensic science administration, which merely investigates the link between error rates in forensic science and the organization of forensic work. Whatever one may believe regarding capital punishment, it remains true that the political opposition to the death penalty has increased the pressure on managers and directors of crime labs to adopt higher standards of professionalism, verifiability, and scientific rigor.

FAILURES OF FORENSIC SCIENCE ADMINISTRATION

There have been many recent documented cases of failed forensic science administration. This fact is connected to the spread of DNA typing, the activities of the Innocence Project, and the politics of the death penalty. The most important of these is probably that of the Houston Police Department Crime Laboratory-DNA/Serology Section. The Houston television station KHOU first reported on trouble in the lab in November 2002 (Raziq and Werner 2004). An FBI Audit of December 2002 found many serious problems including a report that "on one occasion the roof leaked such that items of evidence came in contact with the water" (FBI Director 2002, p. 21). Shortly after this audit, the Houston police discontinued DNA testing at the lab. This case drew considerable attention in the popular press, especially in connection with the voided conviction Josiah Sutton, who was convicted principally on the basis of now discredited DNA evidence from the Houston lab.

Other important cases include the Whitehurst scandal at the FBI crime lab in the late 1990s (Office of the Inspector General 1997), and the current cases surrounding crime labs in Cleveland (Schultz 2004) and Seattle (Teichroeb 13 March 2004), and the fingerprint unit of the Boston Police Department. Cleveland was threatened with a $10 million suit over the wrongful conviction of Anthony Michael Green, who settled for a smaller sum only on the promise that Cleveland would investigate all cases involving test results or testimony from the lab technician who had given false testimony in Green's case (Schultz 2004). The events in Cleveland illustrate the claim that failures of forensic science administration can expose local and state governments to expensive litigation and the possible voiding of large numbers of prior convictions.

These cases and others like them seem to be attracting increasing public attention. Such negative attention may reduce the prestige and perceived reliability of forensic science. They also suggest that error rates in forensic science may be needlessly high. Both the loss of reputation and needlessly high error rates have the same consequence, namely, a tendency to reduce the actual and perceived value that forensic science adds to the American justice system. Here we have a case in which perception is reality in at least some degree. The perception that forensics is unreliable weakens its evidentiary force, which, in turn, reduces the real value it really adds to the judicial system.

CSI Effect

Finally, the popularity of television shows such as CSI has led jurors to have high and sometimes unrealistic expectations of forensic science. The force of the CSI effect seems open to question. Once the effect has been recognized, it may be relatively easy to educate jurors on the nature and limits of forensic evidence. However important the CSI effect may really be, it illustrates the point that forensic science must respond to the external environment within which it operates, regardless of how reasonable or unreasonable that external environment might be.

WHY IT MATTERS

The new environment of forensics matters. Newspaper stories of "faulty forensics" abound (McRobers et al., 2004), defense attorneys are learning new strategies for questioning forensic evidence (Giannelli 2003, Saks 2003), political organizations such as the Innocence Project are publicizing (real or imagined) flaws in the system (http://www.innocenceproject.org), and jurors are demanding higher standards of performance. These environmental factors put pressure on lab managers to adopt new and higher standards of professionalism, verifiability, and scientific rigor.

The forensics community must meet the challenge of these environmental changes. Action on several fronts is required to meet this challenge. We need research, education, outreach, and policy espousal. First, we need research on the principles of forensic science administration. Second, we need to create a separate academic discipline of forensic science administration. Just as we have the distinct academic discipline of public administration, we need a distinct academic discipline of forensic science administration. Third, we need outreach programs through which criminal justice practitioners could keep abreast of current developments in forensic science administration. Fourth and finally, we need policy espousal; we need experts in forensic science administration who will propose new laws, regulations, and government policies aimed at continuous improvement of forensic science administration. Such research, education, outreach and policy espousal would help the forensic science community meet the challenges facing forensic science today.

Researching the Principles of Forensic Science Administration

Research is the first and most important task for forensic science administration. We need to study how the organization of forensic work influences the error rate in forensic analysis. Such research could take several forms. While no one before us has called fore a separate discipline of forensic science administration (with the partial exception of Koppl 2005), there is a heterogeneous body of past work that addresses the central problem at issue. Risinger et al. (2002) apply existing results from psychology to the problem of reducing bias in forensic analysis.

Koppl (2005) applies economic reasoning to many problems in forensic analysis. William C. Thompson (1997) has called for adopting "A Sociological Perspective on the Science of Forensic DNA Testing." Simon Cole (2003, 2005) has produced a series of works based in "science studies," which attempt to cast doubt on the reliability of fingerprint evidence. (Cole's website at the University of California, Irvine says he "specializes in the historical and sociological study of the interaction between science, technology, law, and criminal justice.") From within forensic science, M. A. Thompson (1974) calls for a suite of institutional changes somewhat similar to the proposals in Koppl (2005).

The economics of expert witnessing is an emerging field closely related to the economics of science. This area produces work similar in spirit and content to the sort of research we are calling for. The leading articles are Froeb and Kobayashi (1996) and Posner (1999), and Feigenbaum & Levy (1993, 1996). Froeb and Kobayashi show that the adversary of our courts may have created a social process in which juries make good, perhaps optimal, judgments even those individual jurors are "naïve and biased." They have thereby established an important principle for forensic science administration. Truth, falsehood, and error all emerge from social processes. We should not rely solely on the intelligence, skill, and integrity of individual experts to deliver the truth reliably. We should instead seek out those social processes that most fully encourage reliable testimony.

The research we need in forensic science administration will be closely related to the field of "science studies." The importance of this field is reflected in the decision of the National Science Foundation to create a separate funding program in "Science & Society", which includes a funding category in "Social Studies of Science, Engineering and Technology." Science studies can be very roughly divided into two traditions, one beginning with Merton (1937) and the other with Mannheim (1929&1931).

The broadly Mertonian tradition might be labeled "conservative" in that it tends to support the view that science is relatively objective and truth-driven. In this tradition, the relative rationality and objectivity of science is not a product of any supposed virtues of individual scientists; it is the product of the social structure of science. Important representatives of this tradition today include Kitcher (1993) and Goldman (1999). Kitcher is explicitly engaged in a rearguard defense of science against its critics. The broadly Mannheimian tradition is "radical" in that it tends to support the view that science is ideological and value driven. Important representatives of this tradition today include Bloor (1976) and Fuller (1988). Butos and Koppl (2003) are Mertonians in the broad sense of the term.

We declare ourselves fully for the broadly Mertonian tradition. Simon Cole seems to work within the broadly Mannheimian tradition. In our judgment, this difference makes a difference. It seems difficult to imagine that outside criticisms of forensic science could produce constructive change if the external critics are generally Mannheimian. On the other hand, useful suggestions for institutional improvement are not likely to emerge from within forensic science unless we take seriously the Mertonian principle that the relative rationality and objectivity of science is the product of the social structure of science.

Finally, we hope for experimental research in forensic science administration. Miller (1987) provides an excellent example. He asked a group of 14 students trained in hair analysis to examine four cases each. "The 14 students met the basic requirements for expert testimony on human hair identification in courts of law". For each student, two cases were presented the usual way. They were given two samples and told that one was from the crime scene and the other from the suspect. The other two cases were presented through a forensic line-up.

The known sample from the imaginary crime scene was compared to "five suspect-known hair samples" (Miller 1987, p. 160). In all 56 cases, there were no true matches. The first group of cases showed an error rate of 30.8%. The second group showed an error rate of 3.8%. Miller conducted a pioneering study. His evidence lineup reduced the error rate by about 90%. Scholars should build on Miller's neglected study by applying recently developed experimental techniques (Friedman and Sunder 1994) to a broad range of questions in forensic science administration. Dror et al. (2005) is a more recent example of the sort of experimental work we endorse.

Vernon Smith and his co-workers have created "Economic Systems Design" (Smith 2003, p. 473-474). In this literature, a scholar or group of scholars design a market, for example a market for the sale and delivery of natural gas through an existing network of pipes and wellheads (McCabe et al. 1991). The design is then tested through experimental economics. Experimental economics is similar to experiment psychology, except that subjects in the economics lab are generally given real money payoffs. In our gas example, they trade imaginary quantities of imaginary natural gas over an imaginary network of pipes and well heads. But they get real money to keep for any profits earned from their trading activity.

The money incentives ensure that their decisions are "real" and not just what they idly imagine they might do in posited situation. The laboratory test may reveal design flaws. If so a new design is created and tested in the economics laboratory. The intial "design is modified in the light of test results, the modifications tested, modified again, retested, and so on," Smith explains. "If the final result is implemented in the field, it certainly undergoes further evolutionary change in the light of practice, and of operational forces not tested in the experiments because they were unknown, or beyond current laboratory technology" (Smith 2003, p. 474). The lab is like a wind tunnel. In economic systems design, the criterion by which the designed system is judged is economic efficiency: was the system able to squeeze

Koppl is currently adapting Economic Systems Design for use in testing some of his ideas on forensics. The main change is in the criterion for a successful design. In economic systems design, the criterion is economic efficiency. Do the actors in the designed system act in ways that squeeze out all waste and allow the maximal benefit possible from our scarce resources? In "Epistemic Systems Design," as we may call it, the criterion of success is not efficiency, but veracity. Do the actors in the designed system act in ways that produce results error-free and reliable as possible? We believe that epistemic systems design has great potential as a research tool in forensic science administration.

Creating an Academic Discipline

We call for the creation of a separate academic discipline of forensic science administration. We now have disciplines in "public administration," "healthcare administration," and "business administration," but not in "forensic science administration." Considering the importance of any society's system of justice and the importance of forensic science in the justice systems of modern societies, it seems only reasonable to call for just a discipline.

Creating such a discipline would serve two primary functions. First, it would provide an institutional context for the research we have called for. Second, it would provide a vehicle for the application of that research to forensic science. In particular, it would provide a way for forensic scientists and the managers and directors of forensics laboratories to receive training in the established principles of forensic science administration. Presumably, the preponderance of such training would be through continuing education vehicles such as seminars and certification programs. Such training would be provided by recognized specialists in forensic science administration. It might be appropriate to consider adding such training to the requirements of university programs in forensic science. Forensic scientists would be educated on the nature of the environment in which forensic science is practiced and the challenges thereby created for forensic science; and they would be given training in how to meet the challenges facing forensic science.

This training would serve several ends. First, training in forensic science administration would serve the needs of individual forensic scientists. Training would be aimed in part at helping forensic scientists prepare for testimony in open court. For example, it would help them defend the scientific integrity of fingerprint analysis in a Daubert hearing.

Second, training in forensic science administration would serve the needs of forensic labs. Training in forensic science administration would help crime labs to adapt scientific protocols to the contingencies of daily practice of high throughput lab work. PCR DNA analysis, for example, is extremely reliable if all scientific protocols are followed scrupulously. These protocols, however, are hard to follow (Teichroeb 22 July 2004). Merely greeting a co-worker may cause contamination of a sample from minute bits of saliva, phlegm, or mucus released when speaking. Forensic science administration can design and test protocols that reflect not only the scientific requirements of PCR DNA analysis, but also the contingencies of daily crime-lab work. Training in forensic science administration will then allow forensic scientists to understand and implement such protocols.

Finally, training in forensic science administration would serve the larger institutional needs of the forensic science community. The challenges facing forensic science require us to reconsider some of our basic institutions, such as accreditation, or within-lab verification of results. As we will discuss in a later section, many principles of forensic science administration are principles for the partial re-design of our institutions. The design and testing of such institutional changes is work in forensic science administration. Training in forensic science administration will then allow forensic scientists to understand and implement such institutional changes. For example, the system-wide implementation of "information hiding" would require institutional changes in forensic science.

In part, the need for continuing education programs in forensic science administration arises from new developments from within forensic science administration. More importantly, perhaps, the external environment of forensic science is rapidly evolving. For both reasons, the forensic science community needs continuing education programs in forensic science administration to keep current in the area.

Policy Espousal

Experts in forensic science administration have a role to play in policy debates. The sort of institutional innovations mentioned above cannot be implemented except through the political process. External critics of forensic science also look to the political process for change and should receive a response from experts in forensic science administration. We should also note in this connection that forensic science administration includes the study of how to improve homeland security by improving the work of analytical groups such as the Terrorist Explosive Device Analysis Center (TEDAC), which the Federal government created in December 2003 (Wade 2004). The structure and function of such centers is both a technical and political issue.

LEADING PRINCIPLES OF FORENSIC SCIENCE ADMINISTRATION

Some leading principles of forensic science administration. In this section we explain them more carefully by reproducing almost exactly the brief overview contained in Koppl (2005).

Koppl claims that there are eight features of the organization of forensic science that needlessly reduce the quality of work performed by forensic scientists.

- *Monopoly*: In most jurisdictions today, including those in the US, each laboratory has a monopoly on the evidence it analyzes. No other lab is likely to examine the same evidence.
- *Dependence*: Forensic labs are often organized under the police and are thus dependent on the police for their budgets.
- *Poor quality control*: Quality control systems tend to be weak. In the US, there are no required programs of accreditation and the principal accrediting agency, the American Society of Crime Lab Directors, is a professional group, not an independent organization.
- *Information sharing*: Forensic scientists are privy to information that may be crucial to a criminal proceeding, but extraneous to the questions put to the forensic scientist.
- No division of labor between forensic analysis and interpretation. The same scientist who, say, performs a test to establish blood type judges whether his test results exclude the police suspect.
- Lack of forensic counsel. Indigent defendants rarely receive aid and counsel from forensic scientists hired to help them. In common law countries this creates an asymmetry whereby the prosecution has forensic counsel and, indeed, sometimes great batteries of forensic specialists, whereas the defense has no forensic counsel and, often, an attorney unable to adequately understand and challenge forensic testimony.
- Lack of competition among forensic counselors. From the absence of forensic counsel for the indigent it follows rather trivially that there is no competition among forensic counselors for their custom.
- Public ownership. Forensic laboratories are publicly owned. In the US often organized under police agencies such as the State police or the FBI.

Each of the features just listed discourages unbiased work of high quality. We review each in turn.

- *Monopoly*: The monopoly position of forensic labs allows them to do sloppy, biased, even fraudulent work. As Koppl (2005) shows, recent history provides quite a few examples of poor work. The case of Ralph Erdman illustrates how careless forensic work can sometimes be. This forensic scientist not only faked many of his results, but even managed somehow to loose a human head from a body he was to examine (Kelly and Wearne 1998, p.13). To describe such work as "sloppy" is an understatement.
- *Dependence*: Dependence creates a pro-prosecution bias. For example, an FBI investigator in the Explosives Unit whom we discussed

earlier, David Williams, was found to have "tailored" his testimony "to the most incriminating result" in two separate trials, namely, that of World Trade Center Bombing of 1993 and that of the Oklahoma City Bombing of 1995. In the Oklahoma bombing case, "Williams repeatedly reached conclusions that incriminated the defendants without a scientific basis and that were not explained in the body of the report" (Office of the Inspector General 1997).

- *Quality control*: Quality control measures tend to be poor, which may easily produce persistently poor work. In Scotland, for example, detective constable Shirley McKie was charged with murder on the basis of a fingerprint identification that has been shown to be false and mistaken. An investigation into the case by the Scottish politician and former criminal defense lawyer Winnie Ewing seems to show that "the system of gathering fingerprints in Scotland by the Scottish Criminal Records Office (SCRO) was less stringent than that used in India" (McBeth, 2004). In other words, the SCRO did not have an effective quality control system. The Houston case provides another and very dramatic example, as does the Seattle case (Teichroeb 13 March 2004).
- *Information sharing*: Information sharing between police investigators and forensic scientists creates the strong possibility of unconscious bias (Risinger et al., 2002). It also helps dishonest scientists to act on any self-conscious biases they may have. The inappropriate sharing of bias-inducing information might be called "information pollution." We have already mentioned the case of FBI examiner David Williams who identified the explosives used in the Oklahoma bombing of 1995 and the World Trade Center bombing of 1993. Williams' judgment in each case was found to have been determined by the extraneous knowledge of what the suspects had previously purchased and not by legitimate forensic means.
- No division of labor between forensic analysis and interpretation. Forensic error may result from a false interpretation of a test that was properly conducted. In Houston, for example, George Rodriquez was convicted of rape largely on forensic testimony in which a legitimate test was given an illegitimate conclusion. An affidavit by several leading forensic scientists demonstrates that Houston pathologist Jim Bolding interpreted his serological work in ways that were inconsistent with established scientific theory. "Jim Bolding's trial testimony... contains egregious misstatements of conventional serology. These statements reveal that either the witness lacked a fundamental understanding of the most basic principles of blood typing analysis or he knowingly gave false testimony to support the State's case against George Rodriguez. His

testimony is completely contrary to generally accepted scientific principles" (Blake et al. 2004). There seems to have been a similar disparity between test and interpretation in several cases in Cook County Illinois (Mills et al. 2004).

- Lack of forensic counsel. The lack of forensic counsel has produced many of the false convictions identified by the Innocence Project in the US and the similar British group, Innocent.
- Lack of competition among forensic counselors. Even if forensic counsel is available, it may not be vigorous or effective in a non-competitive environment.
- Public ownership. After the DNA work of the Houston Crime Lab (in Texas) was shown to be unreliable, the Houston Police Department began sending all of its DNA work to private labs. This episode, while merely suggestive, nicely illustrates the claim that private labs may have stronger incentives to produce reliable work than do government labs.

The elements of "competitive self-regulation," Koppl's suggestion for revising the institutional setting of forensic science, consist in reversing the eight characteristics of the current system that we have noted (Koppl 2005). Each element is also a leading principle of forensic science administration.

- Rivalrous redundancy should replace monopoly. There should be several competing forensic labs in any jurisdiction. Subject to the constraints of feasibility, some evidence should be chosen at random for duplicate testing at other labs. The same DNA evidence, for example, might be sent to more than one lab for analysis. The forensic worker need not know whether the evidence is examined by another lab. He will know that there could be another lab, and sometimes is.
- Independence should replace dependence. Rivalrous redundancy and privatization would necessarily create independence in at least the formal sense. Competitive self regulation would produce both formal and substantive independence.
- Statistical review would support improved quality control. For example, if a given lab produces an unusually large number of inconclusive findings, its procedures and practices should be examined. Competitive self regulation creates checks and balances.
- An Evidence Control Officer would substitute information hiding for information sharing. Evidence Control Officer would prepare evidence for testing and shield the lab doing a test from all extraneous knowledge of the case particulars. The Evidence Control Officer should use random-number generators to decide which lab gets a given piece of evidence and when to send the same evidence to more than one lab.

- A division of labor between forensic analysis and interpretation should be applied. When this measure is combined with the provision of forensic counsel for the defense, errors of interpretation are less likely to go unchallenged.
- Forensic counsel should be provided for indigent defendants in criminal cases.
- A system of forensic vouchers for indigent defendants would give forensic counselors to the indigent an incentive to provide high-quality services to their clients.
- Privatization should replace public ownership. Private enterprises are subject to civil liability and may be subject to administrative fines for poor performance. They have, therefore, stronger financial incentives than publicly owned enterprises to provide good and reliable work.

The eight elements of competitive self regulation do not exhaust the principles of forensic science administration. We will mention only one other by way of illustration: scientific opinions are not investigative opinions. An investigative opinion uses all available evidence to judge the probability of guilt. A scientific opinion is based only on scientific evidence and addresses only narrow issues such as whether blood at the crime scene came from a given person. Forensic scientists exceed their authority and put their reputations at risk when they offer investigative opinions in the guise of scientific opinions.

11

Polymerase Chain Reaction (PCR) and Electrophoresis

INTRODUCTION

AIM

The main aim of this case study is to provide a tried and tested polymerase chain reaction (PCR) practical. The premise is to take a sample of plant tissue and get the students to run through three separate PCR reactions: one for 20 cycles, the next for 25 cycles and the last for 30 cycles. Each reaction will use the same starting DNA and the same primers. By running the products of each PCR on an electrophoresis gel it will be possible to see that the fragments of newly synthesised DNA will all be the same size and the amount of DNA generated will be dependent on the number of cycles.

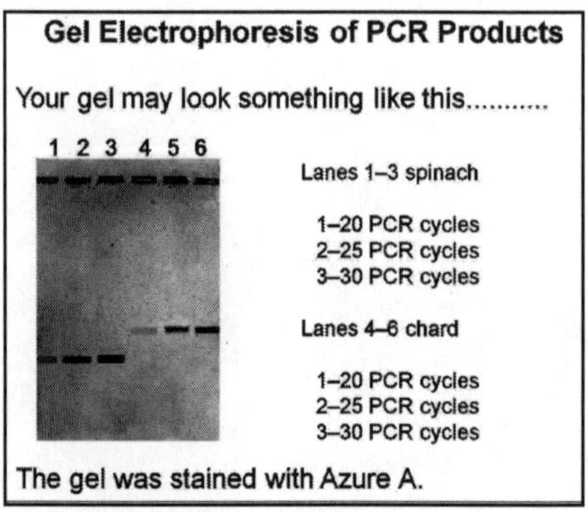

Fig. Slide 21 of the Purification, PCR and electrophoresis PowerPoint

There are two PowerPoint presentations along with work cards that will take the students through the practical. To support the theory behind PCR there is a student's information card, a work card on the basics of PCR and a problem-solving exercise. Links are provided to animations explaining PCR and gel electrophoresis which can be used with students. There is a section on the applications of PCR: links are provided which will take students through the stages of building a DNA profile. Students will get the chance to research how PCR solved a real-life murder case. They will find out how PCR can be used in a number of different scenarios to confirm identity. Finally they will learn how PCR can be used to diagnose genetic diseases.

How to use the case study

First, familiarise yourself and the students with PCR and electrophoresis. Use the animations suggested on page 8 (Setting the scene). These can be viewed via a data projector. You can also use the information card on pages 29 and 30 along with the work card, Basics of PCR on page 27. The work card could be used during gaps in the practical, ie waiting for the gel to run.

Second, decide what you want for your students. If resources and time are available you may wish to do the main practical. However, there is a simple low-cost electrophoresis practical involving food dyes which, although not PCR, may allow the students to simulate DNA profiles. There are also a number of web-based activities. Some are straightforward, others allow students flexibility in researching how PCR is used in confirming identity.

POSSIBLE PRACTICAL ACTIVITIES

PCR: the main practical

The main parts include extracting the DNA, purifying the DNA, running the PCR and gel electrophoresis. At the end of each section you may wish to review what has been achieved so far and the intentions for the next stage. The best place to use the student work card on the basics of PCR is either during the PCR stage or whilst the electrophoresis gel is running.

Alternative practicals

Note there is a relatively low-cost PCR simulation kit available from a commercial supplier. In addition, there is also a low-cost protocol for simulating DNA fingerprints using food dyes. Although not directly related to PCR there are a number of DNA extraction procedures which could be used as stand-alone activities or tied in with restriction digests and electrophoresis.

Problem-solving exercise

Background information on how PCR has been adapted to profile individuals. There is also a short exercise on the applications of PCR in forensic and paternity analysis.

Web-based research activities

There are four activities that have been designed to build on the students' knowledge about the applications and importance of PCR (pages 33–36), some of which offer a degree of personalisation and choice.

SETTING THE SCENE: HOW PCR WORKS

There are many animations and video clips about PCR and electrophoresis available on the internet. The Dolan DNA Learning Centre has two animations on PCR and one on electrophoresis. These would be a good starting point when teaching this topic.

PRACTICAL ACTIVITIES

There are a number of PCR practical activities which are suitable for use in schools. Perhaps the one most people will be familiar with is the National Centre for Biotechnology Education (NCBE) activity, Investigating plant evolution.

NCBE also supply a kit for this investigation containing enough materials for 16 reactions (in autumn 2010 the kit cost £140). They also supply component parts from the kit, which may mean you only need to buy the necessary consumables thus keeping costs down (both primers cost £24 and 17 PCR beads cost £36, as of Spring 2011). The practical activity outlined in this case study is based on the PCR protocol used in Investigating plant evolution.

The aim of the investigation is to amplify regions of chloroplast DNA from different plant species to reveal genetic differences as seen by the difference in size of PCR fragments. However, the aim of this case study is to provide a practical activity on PCR: rather than comparing the differences in chloroplast DNA between species, it would be valid to use the same plant DNA subjected to different numbers of PCR cycles to show that increasing the number of cycles increases the amount of DNA synthesised. This rationale emphasises that lots of specific target DNA can be synthesised from very little starting material. In addition, this is an alternative to analysing more than one type of plant material for comparative analysis, thus dispensing with the need for a DNA sizing ladder and multiple DNA extractions, therefore keeping costs down (see Issue 233 of the SSERC Bulletin, 2010).

The main practical

These notes be read in conjunction with NCBE's Investigating plant evolution student's guide and teacher's notes. Make sure you use the accompanying PowerPoint presentations in conjunction with the student work cards.

There are three stages involved:

1. Extract the DNA.
2. Purify the DNA, followed by PCR
3. Separate the products of the PCR using gel electrophoresis and gel staining.

The DNA sample is obtained by squashing a plant leaf onto an FTA card. This step could be carried out by a technician and the card stored indefinitely and used for several classes over a number of years.

Small pieces of the card containing the plant material are punched from the main card and subjected to several 'cleaning' washes, leaving the purified DNA stuck to the card, ready for PCR and then electrophoresis and gel staining. It should be noted that at the end of each step the procedure can be put on hold by freezing the sample and coming back to it at a later date.

If technician time is available there is no reason why the DNA samples cannot be washed in advance, leaving the students to carry out the PCR and electrophoresis.

EXTRACTION OF DNA

FTA cards

The method outlined in the NCBE protocol uses FTA cards. The FTA card contains SDS and TE (Tris-EDTA) buffer. When a plant sample is pressed onto the FTA matrix, moisture in the sample activates chemicals in the FTA card and these chemicals lyse the cells, denature enzymes, inactivate pathogens and immobilise the DNA. Once completely dry, the plant extracts can be stored for several years at room temperature. The NCBE student's guide states that this method is not suitable for fibrous leaves such as cabbage or pine, which are difficult to crush. The DNA Extraction PowerPoint takes you through this process. Use this in conjunction with the student work card DNA Extraction from Plant Material.

Purification of DNA-impregnated FTA cards

For amplification of the DNA in the plant extract, only a small quantity of material is required and this is obtained by punching a 2-mm disc from the dry FTA card. The disc containing the extract must then be washed twice with purification reagent (to remove inhibitors of the PCR reaction) and rinsed with dilute TE buffer at pH 8. The plant DNA remains bound to the matrix and this provides the template for the PCR. Slides 3 to 9 of the Purification, PCR and Electrophoresis PowerPoint show this process. These slides should be used in conjunction with the student work card Purification of the Extracted DNA.

RUNNING THE PCR

The class may be divided to work in groups. One group will carry out 20 cycles, the second group 25 cycles and the third group 30 cycles. Since students are unlikely to carry out PCR more than once in their school career it could be argued that using three temperature-controlled water baths would give a better learning experience, especially if the three steps of denaturation, annealing and extension were emphasised at the appropriate moment of swapping between baths. On the other hand, a PCR machine is more state of the art and will avoid mistakes, whilst freeing up valuable teaching time – the supplementary work card or extension activities can be used at this point.

Oligonucleotide primers, primer 1 (CHc) and primer 2 (CHd), and a small quantity of water are added to a PCR bead in a PCR tube (the PCR tube is a much smaller thin-walled tube that allows efficient heat transfer). The bead contains the Taq polymerase, dNTPs, buffers and Mg^{2+} salts at the concentrations necessary for amplification of the chloroplast DNA (cpDNA). When the bead has dissolved, the solution is centrifuged briefly to ensure it is well mixed. The 2-mm disc (cut from the FTA card in the previous step) is added, ensuring that it is submerged in the liquid. The tube is then subjected to the amplification procedure. This can be carried out either manually using a series of three water baths or automatically in a thermal cycler. The temperature of the mixture is first raised rapidly to 94°C and held there for 2 minutes to ensure maximum separation of the strands.

This is followed by the predetermined number of cycles (20, 25 or 30) whereby the temperature is held at 94°C for 30s, 55°C for 30s and 72°C for 45s, with the temperature change being executed rapidly. Finally, the temperature is held at 72°C for 2 minutes to ensure completion of DNA synthesis. Since the primers are specific for cpDNA, only the region of cpDNA which lies between the primers is amplified. On completion of the amplification procedure, the PCR tube and its contents can be refrigerated or frozen until required. Slides 10 to 13 of the Purification, PCR and Electrophoresis PowerPoint take show this process. Use this with the student work card Amplification of Chloroplast DNA (cpDNA).

Gel Electrophoresis

Loading the samples can be tricky so it would be prudent to demonstrate how this is done and to let the students practice before the real thing. A blank gel may be reused several times by carefully flushing the wells with a Pasteur pipette between practice shots. Make sure elbows are placed on the bench and both hands are used to steady the pipette: a forefinger can be used to guide the end of the pipette.

By getting three groups of students to work together where each group carries out a different number of amplification cycles, ie 20, 25 or 30, it should be possible to use one gel per three groups. The PCR product is mixed with loading dye and transferred to a well in a 1.5% agarose gel, starting with 20 cycles, followed by 25 cycles and finishing with 30 cycles.

The gel is typically run at 80 V for 30 minutes. It is then stained with Azure A for 4 minutes and de-stained. Since the same plant DNA is used for each PCR the resulting fragments should all be the same size but the intensity of the staining should increase as the number of amplification cycles increases.

Slides 14 to 17 of the Purification, PCR and Electrophoresis PowerPoint show the process for loading the gel. Slides 18 to 20 of this PowerPoint show how to stain the gel. Use these slides in conjunction with the student work card Gel Electrophoresis of PCR Products.

Slide 21 shows what a stained gel looks like, whilst slides 1 and 2 give the aim of the practical and a brief background to the primers used in this case study.

Negative controls

It is good scientific practice to include appropriate controls. However, to minimise costs these can be omitted, although it may be worth discussing them with your class. To test for carry-over of DNA during the extraction procedure, a negative control should be included so, in addition to the disc from the plant species, take a disc from an area of the card that does not have sample applied to it. Treat this disc as above, washing it, etc, as though it were a disc with plant material.

Ideally, each person who sets up a PCR should carry out such a test of their technique. If resources permit, it may be worthwhile to run one negative control per card as an example of good practice. In addition, there should also be a control to test that none of the PCR reagents are contaminated with chloroplast DNA, ie a tube should be set up with the two primers and water but with no paper disc. It can be assumed, however, that all reagents supplied are free from contamination and therefore this second control is not required.

Safety guidelines, equipment and materials

Although the practical may be thought of as being very safe, you should still make yourself familiar with the information in the NCBE teacher's guide and be aware of the precautions and detailed information on all aspects of running the practical that it contains.

ALTERNATIVE METHODS FOR EXTRACTING DNA

NCBE also provides a method of extracting relatively pure DNA from frozen peas or fish sperm: the DNA extracted by this method is good enough for electrophoresis. This is a low-cost method that is very similar to the classic DNA your onions method, which can also be found on the NCBE website. Both these documents come with detailed equipment lists, procedures and safety precautions.

Traditional method for extracting plant DNA (more laborious)

A small disc of plant material is homogenised with sand, SDS, Tris at pH 8, NaCl and EDTA (two to three volumes per gram of tissue). Sand helps to break open the tough plant cell walls. SDS is a detergent that helps to break open cell membranes to release the DNA. Tris is a buffer that maintains the pH for DNA extraction at around 8. NaCl also helps to break open the cells and also helps to precipitate DNA. EDTA chelates Mg2+ ions, which helps to break up protein complexes. Solid debris is removed by centrifugation and the supernatant, which contains the DNA (genomic and cpDNA), is removed to a clean micro tube and the DNA precipitated by addition of ice-cold ethanol. The mixture is again centrifuged to pellet the DNA and the supernatant removed. The DNA (still a mixture of genomic and cpDNA) is then resuspended in sterile water.

SOURCES FOR OTHER SUITABLE PCR EXPERIMENTS

SIBE resources: green fingerprinting scenarios

Although these protocols do not involve PCR, the likelihood is that in the world of research PCR would be the preferred strategy because it requires less high-quality DNA, and is quicker and less expensive. These protocols are included here as a matter of interest.

Simulating DNA fingerprints using food dyes

The science of making a DNA profile has advanced a great deal since its invention in 1984. PCR has largely superseded the more traditional method, which relies on having a substantial amount of high-quality DNA, several restriction enzymes, Southern blotting apparatus, labelled DNA probes, X-ray film and a developer or suitable alternative such as phosphorescent screens. The PCR and Forensic Analysis: Problem Solving exercise (page 32) briefly outlines the FBI's current method of generating a DNA profile.

The SAPS Biotechnology Scotland Project, in their practical The Wonderful Wizardry of Finding a Gene, use mixtures of common food dyes separated by gel electrophoresis to reveal the profiles of different coloured bands:

'This practical uses standard agar and water instead of the more expensive agarose and buffer, and mixtures of food colourings instead of DNA. Each of the four food colouring mixtures represents chromosome extracts from different wizards and has a green appearance. When the mixtures are subjected to gel electrophoresis, each separates into a different pattern of coloured bands. The different coloured bands indicate the 'magic power genes' possessed by each wizard.'

Gels can be made in advance for the students. Each gel will require about 16 ml of molten agar:

1. Make up a 3% solution of agar in a flask and place it in a boiling water bath until the agar melts and the solution goes clear (place clingfilm over the flask to prevent excess evaporation).
2. Cool the agar to 60?C.
3. Meanwhile slot the combs into electrophoresis tanks and place on a level surface.
4. Pour the molten agar into the centre of each tank so that it flows between the teeth of the comb. The agar should be about 5 mm thick (so that its surface is level with the plastic ridges which form

the end channels). Try not to pour agar into these end channels (if it does go into the channels it can be scooped out later when the agar has set).

5. Leave the gels to set and then remove the combs.
6. Cover the gels with water to stop them from drying out. Top up the water level, if required, when the gels are to be used.

The 'DNA' samples are made from liquid food colourings that have had sucrose added to them to help the samples sink into the wells.

1. Add 3 g sucrose to every 5 ml of dye required and dissolve (for 10 groups you will require a total of 12.5 ml of green food dye, 5 ml of blue, 2.5 ml of yellow and 1.0 ml of black).
2. Make up samples in labelled tubes as follows:

Sample	Food Dyes*
1	Green
2	1 blue:1 yellow
3	4 green:1 black
4	1 blue:1 green

Note:

*The colours may vary depending on the brand or shade of food colouring.

3. Aliquot 0.5 ml samples into small labelled tubes for the students. The gels are loaded and run in the normal way (see the student work card Gel Electrophoresis of PCR Products) and typically take less than 30 minutes to run. Moreover, the gels do not require staining, giving instant results.

Technicians' guide

DNA extraction (per group or per class – each group/person requires one 2-mm disc containing the DNA)

- 1 FTA card, which can hold enough DNA for several classes
- Plant samples (spinach or red chard work well)
- Pestle
- Backing board
- 2-mm punch

Decide if the students will extract the DNA or it will be prepared in advance. Materials (per pair or individual)

- Marker pen
- Micropipettes (P20 and P200 or suitable alternatives – see the NCBE student guide for fixed volume pipettes and advice on graduated pipette tips that can be use with a plastic 1-ml syringe)
- 1.5 ml micro centrifuge tube
- Box of yellow pipette tips (approx 15 tips)
- Purification reagent

- TE-1 buffer (10 mM Tris HCl, 0.1 mM Na2EDTA)
- Oligonucleotide primers (10 µl each, CHc and CHd – see the NCBE student guide)
- 1 PCR tube containing a PCR bead (the PCR bead contains Taq polymerase, dNTPs, buffers and Mg2+ salts)
- Micro tube stand (a piece of polystyrene with suitable holes cut into it works well)
- Loading dye
- Cocktail sticks
- Azure A stain
- Micro tube containing DNA ladder (optional)
- Electrophoresis tank with 1.5% agarose gel (see the NCBE student's guide)

Materials (per class)

- Thermal cycler or
- Stop clock and three water baths:
  - One set at 55°C
  - One set at 72°C
  - One set at 94°C (this should be covered to maintain the high temperature)

Student work cards

DNA Extraction from Plant Material:

1. Place a backing board between the back cover of the card and the absorbent layer. Close the cover over the plant material.
2. Fold back the cover of the FTA® plant card. Take a fresh leaf or piece of plant material and place it onto the extraction card. Ensure that the leaf does not extend outwith the box on the printed card.
   Note: Spinach and red chard work well.
3. Using a pestle, squash the leaf onto the card until the moisture has soaked through to the back of the card. Try not to let the sample spread outwith the box as this can contaminate adjacent samples.
4. Fold back the cover and discard the plant material.
5. Label the appropriate sample ID box on the front of the FTA® plant card with the plant name.
   Note: The card should be labelled after the plant material has been squashed onto the card as the writing can come off during extraction/ squashing.
6. With the cover folded back, allow the card to air dry at room temperature for a minimum of 1 hour.
   Important: Do not heat the card as this may fix inhibitors of the PCR onto the card matrix.

Once dried, the card can be sealed in a plastic bag with a sachet of desiccant and stored indefinitely.

PURIFICATION OF THE EXTRACTED DNA

At the end of this section you will have a small disk of card with the purified DNA stuck to it:

1. Place a clean cutting board between the dry FTA® plant card and the backing. Place the tip of the 2-mm punch over the area to be sampled, ie an area within the box coloured green by the chlorophyll. Try to select areas which have the same colour intensity. Press firmly on the punch and rotate to remove a paper disc.
   Note: To avoid cross-contamination, do not take a disc close to the edge of the box if the extracted sample has overlapped with another sample in an adjacent box.
2. Using a cocktail stick, transfer the small disc into a clear 1.5-cm3 micro centrifuge tube. Label the tube with the species used.
   Important: If the punch is to be used for a second sample, first clean the punch by removing a disc of paper from an area of the card that does not have plant material applied to it. Discard this disc. This prevents contamination of subsequent samples.
3. Using a micropipette, add 150 μl of purification reagent to the tube containing the paper disc.
4. Close the tube and flick the bottom to wash the disc. Flick every 30 seconds for 2 minutes. This removes PCR inhibitors from the disc.
   Important: If the tube is not agitated, the disc will not be washed and the PCR may fail.
5. Remove and discard the purification reagent from the tube. Remove as much of the froth from the tube as you can.
6. Repeat steps 3 to 5.
7. Using a new yellow tip, add 150 μl of TE-1 buffer to the tube.
8. Close the tube and flick it to wash the disc. Flick every 30 seconds for 2 minutes.
   Note: This removes SDS (detergent) from the FTA® disc.
9. Remove and discard the TE-1 buffer from the tube.
10. Repeat steps 7 to 9.
11. Go on to conduct the PCR. Alternatively, the disc can be dried in the tube by warming to approximately 50°C. Once dry, place the closed tube in a sealed foil bag with a desiccant sachet and store at room temperature until required.

Amplification of Chloroplast DNA (cpDNA)

The aim of this experiment is to use three similar DNA samples: treat one to 20 cycles of PCR, one to 25 cycles of PCR and the third to 30 cycles of PCR. After completion of the PCR amplifications each sample will be analysed by gel electrophoresis. The three samples will be run side by side for direct comparison (your teacher will help you get into groups).

1. Label a PCR tube containing a PCR bead with your initials. Label both the lid and sides of the PCR tube clearly.
2. Add 4 μl of sterile deionised water to the PCR tube.
3. Using a fresh tip, add 10μl of primer 1 (CHc) primer suspension to the PCR tube.
4. Using a fresh tip, add 10μl of primer 2 (CHd) primer suspension to the PCR tube.

   Note: Ensure that the tip does not touch the PCR bead. If it does, the bead may stick to the tip. Instead, place the tip against the side of the tube above the bead. This will leave a drop of liquid on the side of the tube. The liquid will run down to the bottom and the PCR bead will dissolve.
5. Close the tube. Flick the bottom of the tube gently to mix the contents and help the PCR bead dissolve.
6. Spin briefly in a micro centrifuge tube to collect the sample in the bottom of the tube.
7. Using a clean cocktail stick, transfer the disc impregnated with plant DNA to the PCR tube containing the PCR reagents.

   Important: Ensure the disc is submerged in the PCR reagents at the bottom of the tube.

   You are now ready to carry out the PCR on your sample.
8. The temperature of the mixture is first raised rapidly to 94°C and held there for 2 minutes to ensure maximum denaturation (separation of the strands).

This is followed by a number of cycles (either 20, 25 or 30)

94°C for 30 seconds	denaturation
55°C for 30 seconds	annealing of primers
72°C for 45 seconds	extension

Finally the temperature is held at 72°C for 2 minutes to ensure completion of DNA synthesis. At end of the cycle collect your PCR tube. Only the cpDNA will have been amplified. This can be frozen at –20°C if it is not required for some time.

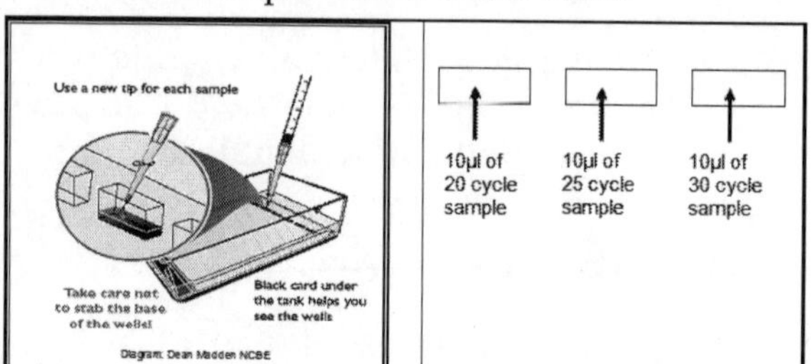

Gel Electrophoresis of PCR Products

1. Using a fresh tip, add 2μl of loading dye to your amplified sample and mix.

The loading dye is much heavier than the electrophoresis buffer and should sink to the bottom of the well. You may want to practise loading a blank gel with loading buffer before using your sample. The wells can be rinsed out using a Pasteur pipette and reused.

2. Load 10 μl of the sample into a well in the gel.
3. Using a clean tip for each sample, load each of the remaining wells with 10 μl of a different PCR product and loading dye, then start the electrophoresis. (Running a DNA ladder is optional.)
4. When the electrophoresis is complete, remove the gel from the tank and place it in a staining tray. Stain the gel with stain for 4 minutes.
5. Pour off the stain (back into the bottle) and then rinse the surface of the gel with water three or four times. Be careful not to leave any water on the surface of the gel. Faint bands will be visible after 10 minutes but the bands will become clearer if left overnight. To prevent the gel from drying out place it in a plastic bag.
6. Record your results and draw a conclusion on the evidence you have generated. How did this compare to what you thought would happen?

Basics of PCR

Use the information card on PCR to help you answer these questions:

1. Whilst setting up your PCR tube you added your DNA sample, some sterile water, the two primers and a PCR bead.
   What do you think is in the PCR bead?

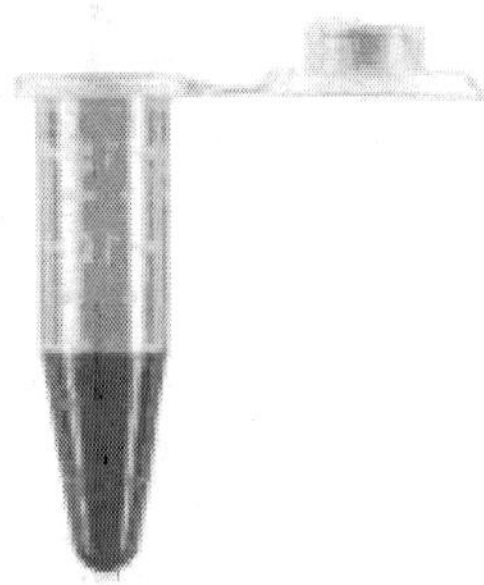

2. Throughout this experiment you have taken precautions, such as only using one type of plant material, using a clean cocktail stick to transfer your DNA sample and using fresh pipette tips whilst dispensing the primers. Why do you think this is?
3. There are three stages to each amplification cycle. What happens during:
   - Denaturation (30 seconds at 94°C)?
   - Annealing (30 seconds at 55°C)?
   - Extension (45 seconds at 72°C)?
4. What feature allows the DNA polymerase, used in PCR, to remain active?

5. At the end of the PCR cycling why was only the target DNA synthesised?
6. Can you think of any controls that should have been run during this experiment?
7. Assuming there was only one target sequence of DNA present in your sample and the reaction was 100% efficient, how many copies would be present after:
   - 20 cycles?
   - 25 cycles?
   - 30 cycles?
8. Predict what you expect to see after you have completed the gel electrophoresis.

Note on chloroplast DNA

The primers used in this study are specific to highly conserved genes found within plant chloroplast. These genes encode transfer RNAs (tRNAs), which have not changed much during the course of evolution, and identical sequences can be found in the chloroplast DNA of most higher plants. However, the DNA which lies between these tRNA genes can be highly variable and is more prone to mutations. The primers have been designed so that the target sequence covers such a highly variable region of chloroplast DNA. This means that any DNA amplified from the same species of plant should be the same sequence and size. However, if two species which are not related are compared then the target sequences will, in all probability, be different in sequence and size.

STUDENT INFORMATION CARD: THE POLYMERASE CHAIN REACTION

From a single piece of DNA, PCR is capable of making billions of copies of a particular sequence. This relies on all the ingredients needed for DNA replication being present: the target sequence, or template DNA, free deoxynucleotides, DNA primers and heat-stable DNA polymerase such as Taq polymerase. Typically the primers are about 20 nucleotides long and are complimentary to the ends of the target sequence.

Usually 30 cycles, or reactions, are carried out one after the other. Each cycle is made up of three parts:

1. *Denaturation*: The mixture is heated to 90°C to separate the two strands of DNA.
2. *Annealing*: The temperature is lowered to 55°C, allowing the primers to specifically bind to the target sequence by complimentary base pairing.
3. *Extension*: By heating to 72°C Taq polymerase will synthesise new DNA from the target sequence.

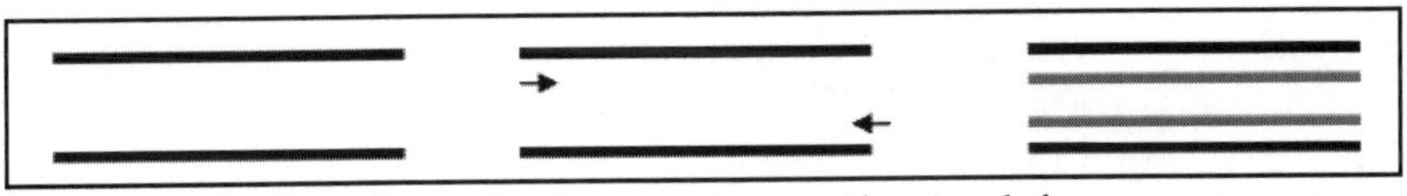

At the end of each cycle the newly synthesised fragments act as fresh templates so if there is a single piece of DNA to begin with then after the first cycle there would be two, after the second cycle four, after the third cycle eight, after the fourth cycle sixteen, and so on (doubles every cycle).

The primers are written starting with the 5' end (phosphate of the first nucleotide) and finishing with the 3' end (deoxyribose of the last nucleotide).

5' (phosphate end) ? CGAAATCGGTAGACGCTACG ?
3' (deoxyribose end)

(primer 1/CHc)

5' (phosphate end) ? GGGGATAGAGGGACTTGAAC ?
3' (deoxyribose end)

(primer 2/CHd)

DNA polymerase will only add nucleotides to the 3' (deoxyribose) end of the primer or to the growing chain of newly synthesised DNA. Only bases which specifically compliment the DNA template will be joined to the strand being synthesised, ensuring that the original DNA sequence is copied letter for letter or base for base.

Each strand of DNA in the double helix runs in opposite directions, ie the strands are anti-parallel. The arrows show the direction of synthesis.

Template DNA	3' (deoxyribose end)	.......**GCTTTAGCCATCTGCGATGC**.............	5' (phosphate end)
Primer	5' (phosphate end)	CGAAATCGGTAGACGCTACG ›	3' (deoxyribose end)
Primer	3' (deoxyribose end)	← CAAGTTCAGGGAGATAGGGG	5' (phosphate end)
Template DNA	5' (phosphate end)	.......**GTTCAAGTCCCTCTATCCCC**.............	3' (deoxyribose end)

By knowing the target sequence it is possible to make billions of copies of a chosen piece of DNA in a relatively short time. Since the primers used in PCR are unique to each target sequence, the PCR reaction is very specific and in theory can amplify a single DNA sequence from a complex mixture of DNA molecules.

PCR is a valuable analytical tool and is routinely used for research purposes; diagnosing diseases, be they inherited or infectious, genetic fingerprinting, paternity cases, forensics, quality assurance in the food industry and even molecular archaeology. However, because of its incredible sensitivity scrupulous precautions have to be taken to keep unwanted DNA out of a reaction mixture.

PCR AND FORENSIC ANALYSIS: PROBLEM SOLVING

Forensic analysis has been revolutionised by PCR. The human genome has regions of repetitive DNA called short tandem repeats (STRs). STRs are typically two to seven bases long and, as the name implies, are repeated –

almost randomly – with each stretch of repeats being unique to an individual. STRs are found throughout the human genome at different locations on the same chromosome and on different chromosomes.

Forensic scientists have harnessed the PCR reaction to analyse STRs from DNA recovered at crime scenes. For example, the FBI uses 13 different STR sites to profile individuals. They do this by combining 13 pairs of unique primers, each pair specific to its own STR site and amplifying the target DNA simultaneously. In other words they carry out 13 PCR reactions at the same time in the same tube. This technique is called multiplex PCR.

The PCR fragments that are generated from the multiplex reaction can be distinguished from one another because the primers are pre-labelled with a fluorescent dye. This colour codes the PCR fragments and allows the scientists to immediately know which STR they are looking at once the PCR fragments have been separated by electrophoresis. Remember, the STRs will vary in size depending on the nature of their tandem repeats.

The picture is complicated because each STR will have two alleles, one derived from the mother and one from the father. So 13 pairs of primers will generate 26 PCR fragments! Since PCR works best when amplifying short stretches of DNA, STR analysis has the advantage that it will be more likely to work on DNA that has been damaged by adverse environmental conditions such as severe decomposition. Also, by analysing many STR sites the chances of misidentifying a suspect are reduced to practically zero.

APPLICATIONS OF PCR

Building a DNA profile

By working through these two web activities you should become familiar with the process of DNA profiling. You may want to read through the PCR and forensic analysis: problem solving exercise first.

Forensic science: the Cardiff three

One of the most amazing uses of PCR involved a murder case, which has become known as the 'Cardiff three'. In 1988 a Cardiff prostitute was brutally murdered by being stabbed over 50 times and having her throat cut. Three men were initially convicted of the murder but were acquitted in 1992, whilst the real killer was still at large.

In completing this research project:

- You will find out how sensitive PCR can be
- You will discover that DNA profiling relies on the use of databases
- You will find out that those who commit crimes but whose information may not be on a database can still be brought to justice.

Make rough notes on the following:

1. When the victim's flat was swept for traces of DNA 12 years after the crime, where did scientists find new traces of DNA?

2. Why was the suspect known as 'cellophane man'?
3. What was the significance of the second-generation multiplex plus PCR test?
4. Once a DNA profile had been found no positive match was found in any databases, so how did the database information yield the killer?

This case highlights how PCR is capable of detecting minute quantities of DNA and how specific the reaction can be.

Write up your notes in the form of an essay:

- You should have an introductory paragraph on the uses of PCR in forensic analysis. It does not have to be about the Cardiff three.
- Your next paragraph or two should include the information about PCR you found out whilst researching the Cardiff three.
- Finally, you should finish with a paragraph summarising the benefits that PCR has brought to this field.

Identification using DNA

On this web page you will come across sections on how the DNA science has aided forensic identification:

- Linking suspects to the scene of a crime
- Securing the release of innocent people from jail
- Identifying missing persons from disaster sites
- Identifying soldiers killed in action
- Confirming the identity of long-dead monarchs
- Ecological tracking
- Novel commercial ventures
- The discussion of ethical issues surrounding DNA databases.

Choose an area which interests you•

Write a paragraph on why this is important and worthy of having money spent on it. Your next section should highlight how the science behind DNA has made this possible, in particular the application of PCR.

Finally, summarise your thoughts on what you have written, highlighting the advantages and disadvantages that the science has brought to the matter.

Index